I0475143

The Progression of Time

of Time

How the expansion of space and time
forms our world and powers the universe

C. Johan Masreliez

ISBN: 1456574345
ISBN 13: 9781456574345

Library of Congress Control Number: 2011901412
CreateSpace Independent Publishing Platform
North Charleston, South Carolina

Table of Contents

Abbreviations:

AT	**Atomic time**
CMB	*Cosmic Microwave Background*
CRF	*Cosmological Reference Frame*
DIST	*Dynamic Incremental Scale Transition*
ET	*Ephemeris Time*
GR	*General Relativity*
JPL	*Jet Propulsion Laboratory*
LT	*Lorentz's Transformation*
QM	*Quantum Mechanics*
QT	*Quantum Theory*
RTE	*Rule of Timeless Existence*
SCM	*Standard Cosmological Model*
SEC	*Scale Expanding Cosmos*
SR	*Special Relativity*
UT	*Universal Time*
VT	*Voigt's Transformation*

ACKNOWLEDGEMENTS

I thank my daughter Dr. Malin Young for her editorial help and suggestions, and my son Carl Jonas for his support and computer wizardry. I also thank Mr. Del Miller for reviewing and providing several helpful suggestions and for the design of the front and back cover. I am also thankful for the continued support over the years given by Mr. Kurt Borgne, a class mate from my early days at the Royal Institute of Technology in Stockholm, and my brother Karl-Gustav who has contributed with many helpful suggestions. Last but not least I am grateful for the steadfast support my wife Anne-Marie has given me over many years.

To Anne-Marie, Malin and Jonas,
Amy, Niklas and Markus,
and to all who will see.

PREFACE

WHAT WE NEED IS NOT THE WILL TO BELIEVE BUT THE WILL
TO FIND OUT.

—BERTRAND RUSSELL

When I began writing this book, I thought I would more or less summarize my findings over seventeen years of independent solitary research, during which I eventually was able to publish some of the results presented in this book. Unfortunately, these published scientific papers are inaccessible to most people, because they demand considerable background in theoretical physics. Realizing that this might be a problem, I decided to abandon the academic approach and present what I have to say in plain language because I think it may be of interest to people in general. The disadvantage of this approach is that my presentation will not be as precise as a scientific treatise, but this is OK and might even be better, because what I am going to tell you is not that difficult to grasp.

This book is mainly about cosmology, which is the scientific treatment of the universe as a whole. In the past people have always been keenly interested in this subject, which is not surprising. We are born into the world without any prior knowledge, and most of us are curious about the universe and want to find out as much as possible about it. All cultures have responded to this desire by creating their own cosmologies and cosmogonies, where cosmogony deals with the origin and evolution of existence.

My interest in cosmology was stirred when I found that all is not well with our current ideas about the universe as represented by the Big Bang theory. By the current thinking the world was created at a single instant, and in a state of infinite energy density, after which it has been expanding ever since, eventually to form the universe we see today. However, many scientists with me find that there are serious problems with this creation story; I will in this book introduce an alternate theory that better agrees with astronomical observations as well as with our common perceptions about the world.

In my opinion the most troubling aspect of modern cosmology is its missing explanation to the progression of time; it does not address the physical process that makes time pass. Obviously, a scientific theory that fails to explain the progression of time, which arguably is the most keenly felt aspect of our existence, cannot be the last word. And as we shall see, knowing what causes the progression of time may be the key to understanding the cosmos.

However, it turned out that suggesting an explanation to what makes time progress became a main difficulty in writing this book, because it requires new physics. The usual approach in science is to gradually add new discoveries while building on past knowledge. But, a problem arises if a new discovery does not fit into this past knowledge, but conflicts with known epistemology. I was faced with this obstacle when beginning to write this book. It became a main difficulty because it meant that I could not approach the solution to the progression of time in the usual way, basing it on known science; instead I had to introduce a new cosmos model that explains the cosmos better than the Big Bang model. The explanation to the progression of time then became an implicit consequence of this new model.

In his book "On the Revolution of the Celestial Spheres" published just before his death in 1543, Copernicus suggested that the Earth is but one of several planets in motion around the Sun. He was careful not to challenge authorities because at the time everyone thought that the Earth was immovable and at the center of the universe. However, in his book he was able to convincingly support his new model by numerous examples, demonstrating how it would explain the motion of the night sky and of the other planets as seen from the Earth as well as their changing appearances with time. Interestingly, Copernicus argued that motion is relative and that the observed motion of celestial objects did not reveal whether the Earth was fixed or in motion.

Similarly I will argue that our current *scale* of space and time (from here on simply denoted "spacetime") may not be fixed but changing with time and that our astronomical observations may reflect this motion in the scale of spacetime. I will shortly return to what I mean by "motion in the scale of spacetime", which appears to be a new concept, not previously considered in science.

Like with the Copernican revolution the new cosmos model I propose implies a huge shift in perspective. From previously having been fixed, Copernicus set the Earth in motion, while similarly the new model sets the whole world in motion in the direction of a steadily increasing cosmological scale of spacetime, a scale that in the past always has been considered fixed. I will demonstrate how this scale expanding cosmos model will solve numerous cosmological puzzles and lead to a new and better understanding of our world.

In this presentation I will ask the reader with strong background in physics to remain patient; the technical justification for the new ideas to be introduced in this book may be found in already existing physics and in my published papers, Some of it is included in the book's appendices. The presentation will gradually move deeper and deeper into the details and may reward the patient reader with a new enlightened world-view.

This book might help us get out of the dead end in which physics now finds itself, and help improve our understanding of the world. And, in doing so, it will revise and reinterpret the Special Relativity theory as well as extend the applicability of General Relativity. This might seem like a tall order, and perhaps you might think these pillars of modern physics should not be questioned. Although it of course is possible that I am wrong, I can only tell you that *if* I am right, this book could change the direction of physics as well as irrevocably revise our worldview.

When reading this book, you will join me on a journey of discovery that uncovers a fundamentally important but previously neglected aspect of our existence.

This work began with a simple idea regarding the nature of the cosmological expansion. However, during the years I have been investigating the consequences of this idea it has gradually become clear to me that a hidden process of fundamental importance exists in the universe that directly influences all aspects of our existence via the progression of time.

The passage of time has always been enigmatic to us; nobody really knows what might cause it.

> WHAT THEN IS TIME? IF NO ONE ASKS ME, I KNOW WHAT IT IS. IF I WISH TO EXPLAIN IT TO HIM WHO ASKS, I DO NOT KNOW.
>
> SAINT AUGUSTINE

Modern physics has no answer either; the theory of General Relativity gives no clue of what might cause time to progress. However, this book proposes a well-defined physical process that could be the essence of "the progression of time." Unknowingly we may be living "on the back of a dragon" that sustains all existence. When its presence is acknowledged, it could shed new light on the world and our human predicament in the cosmos. It could have great impact on all aspects of life here on Earth and on the future space exploration.

But let's start from the beginning.

When we think of empty space, we might imagine a void in which nothing exists. But, since things obviously exist in this world, we might ask what is determining the scale of things. In other words, what could in a perfect void possibly determine the scale of, for example, an apple if there is nothing to compare it to? In the past, this question has not arisen because people believed that God created the world and implicitly also decided its scale. But, if we don't accept this explanation and instead believe that the cosmos is all there is, the universe must somehow decide the scale of material objects via some kind of internal property or mechanism.

This insight made me investigate the implication of changing the four-dimensional scale of things so that all distances and time intervals change by the same scale factor. It turns out that worlds of differing scales would seem identical for inhabitants with matching scales.

In other words, the scale of space and time is a free parameter that does not influence physical processes or our perception of the world.

Since this point is of crucial importance for the development to follow permit me to elaborate a bit. All material objects including our bodies are made up of atoms of various kinds. Each atom takes up a certain volume in space, and the volume and size of any object is determined by the combined volumes of all its individual atoms. However, each atom is not simply a dead volume but atoms are dynamic objects with electrons whirling about their nuclei. They are oscillating entities that have found use in atomic time, which presently is used extensively in super accurate clocks. Therefore, the properties of all atoms are determined not only by their volumes but also by their pace of time.

We might say that the scale of space and time determines the scale and size of matter. However, it turns out that this scale is not predetermined but that all aspects of our world remain the same if its scale were to change.

It is therefore possible that no fixed scale exists in the universe, but that the scale actually is steadily increasing. This would immediately resolve the most puzzling problem of ancient as well as modern cosmology, because it would explain the origin of the world. Or rather, if the cosmological scale were to increase by a tiny fraction every second, it would allow perpetual existence. As we shall see, an incrementally increasing scale would also explain the origin of the energy that permeates the cosmos and keeps it going.

With increasing astonishment and awe I investigated this new Scale Expanding Cosmos (SEC) model and found that it would perfectly explain the world we live in.

A great number of books and articles are now available on modern cosmology, covering subjects like the Big Bang creation event, cosmological inflation, dark energy, dark matter, black holes, accelerating cosmological expansion, and so on. Some of these publications also mention unresolved conceptual puzzles such as the mysterious origin of the universe or what might cause the progression of time. There have also been several TV programs describing how the universe began in the Big Bang event, expanding out of an infinitely small "singularity," where the laws of physics as we currently know them did not apply. According to this story of creation, the Big Bang was followed by cosmological expansion, which eventually, after some 14 billion years, has formed the universe we see today.

However, these well-staged presentations usually do not mention that there are actually many problems with this popular explanation of the universe. The most obvious mystery is undoubtedly the creation event, in which all there is supposedly was created from nothingness. There are numerous additional problems with the Standard Cosmological Model, a theory that in the following will be referred to as the SCM. These difficulties are further discussed in Chapter 4 of this book. Here in the Preface, I will only mention a few of the most glaring problems and puzzles to set the stage for the new and better cosmos model to be proposed and justified in this book.

The most obvious problem with the SCM, besides the creation event, is that it offers no explanation to what might be causing the progression of time. The theory is based on

General Relativity (GR), but *GR is a theory that cannot model the progression of time*. GR deals with four-dimensional (4D) geometry. It describes a static world in which the four dimensions of space and time (which in the following is denoted "spacetime") are fixed and do not change. According to GR, ordinary motion in space is modeled as a "world-line" in 4D spacetime, and every point on this world-line corresponds to a certain location in space and time. But nothing in GR describes the *process* of moving along such a world-line of existence. Consequently, in GR there is no difference between the past, the present, and the future; GR gives a *static* picture of a 4D cosmos where all epochs coexist.

Thus, the GR theory is mute when it comes to describing the physical process, which we all experience as the progression of time. Strangely, this has led some people in science, including Albert Einstein, to question whether time really progresses in a scientific sense.

Obviously, a theory of the cosmos that cannot explain the progression of time must fall short.

If science cannot explain something as fundamental as the progression of time, something important must be missing. In fact, it seems that the progression of time should be of central importance to any model of the cosmos. In this book, it will become clear that this actually is the case; there is a new cosmos model in which the progression of time is crucial.

Ignoring the progression of time could perhaps be justifiable in Einstein's first model of the universe of 1917, since he used GR to describe a static cosmos that never changes [Einstein, 1917]. However, it must be considered strange that the SCM by which the universe expands does not provide an explanation to the progression of time.

Many have been puzzled by the nature of time. Einstein speculated that perhaps its progression is an illusion. Although we all keenly experience it, it might be something that simply cannot be explained by physics. Unfortunately, this would make a most important aspect of existence something "beyond science." As we shall see, Einstein was right in that the progression of time cannot be modeled by known physics. Ironically, GR needs modification before this may be done.

The failure to understand what is causing the progression of time has led to numerous problems not only with our cosmos model but also with modern physics in general. It turns out that it is intimately related to Quantum Theory (QT), and that failure to understand what is causing it makes GR incompatible with QT.

There are several additional unexplained problems with the SCM. For example:

- Dark energy
- Dark matter
- Entropy
- The horizon and flatness problem
- Accelerating cosmological expansion
- Cosmological inflation
- Cosmological evolution
- Observational discrepancies

These are further discussed in Chapter 4. However, the resolution to all these puzzles will require the use of new physics to be presented later in this book.

In this book I will address these problems and show how the new cosmos model resolves all of them.

However, from what you read in books and articles and see on TV, you might be unaware of all these problems with the SCM because typically they are swept under the rug in these well-staged productions. But readers who have become aware of the shortcomings of the SCM might feel a bit confused and start suspecting that a consistent and conceptually satisfying explanation to our world does not yet exist. This book rectifies this situation by demonstrating that there actually is a better model.

Although the Big Bang idea has received much attention from the academic community and from specialists working on cosmology (cosmologists), many ordinary people (laymen) wonder about the creation "of everything from nothing" since it is hard to understand how anything at all can have been created from nothingness. And, science does not offer any reasonable explanation to this miracle. We accept the Big Bang event because everyone else does, even though it really makes no sense. Intuitively, the Big Bang idea seems very strange and perhaps even wrong. And speculation of "mother universes" spawning new "baby universes" seems far-fetched to say the least. Since this may never be confirmed by observation or measurement, it does not belong to science.

One might wonder how the scientific Big Bang creation idea arose to begin with. Of course, the belief that the universe somehow was created has had a long history in human culture, often in the context of religion. But how did it become a scientific theory?

In order to explain this, let us review what is meant by knowledge, in particular scientific knowledge. Many believe that science consists of well-established, widely accepted epistemology; people believe that science is what may be found in books and in articles published in mainstream journals. The knowledge that constitutes scientific epistemology is used as a guideline when assessing the validity of new ideas.

Modern cosmology is, in essence, physics as applied to the universe. Thus, it interprets our world by making use of available knowledge like Special Relativity, General Relativity, Thermodynamics, Quantum Theory, and so on. Consequently, a scientific explanation of our world must use ideas and concepts conforming to this accepted knowledge base; explanations using ideas not found there typically are considered unscientific, and their publications might be, and often are, obstructed.

However, remembering our past, with worldviews that nowadays are considered hopelessly outdated, we should realize that at the time when these now-obsolete worldviews were in vogue, they represented the current understanding; they reflected the knowledge of their time. The Flat Earth – and the Ptolemaic Earth-centered worldviews were, at the time, eminently believable; the only reason why we eventually discarded them was that better explanations became available.

Since the scientific community relies on established knowledge when judging new ideas, even a strange idea like the Big Bang creation may find credence if it can be accommodated by familiar concepts and theories. In other words, an idea that intuitively seems strange might find acceptance if it can be described by known physics. Since a spatially expanding universe may be modeled by GR (if we ignore the problematic progression of time), it is used to model the SCM, which at least partly explains the general acceptance of this cosmos model.

However, we should always keep in mind that important knowledge might be missing that prevents us from finding a better cosmological model than the SCM. We must acknowledge that what we currently know is but a miniscule part of what mankind will know in the future. Even if we are aware of the fact that we really don't know much yet, it seems that we as humans always try to form a worldview based on what we currently know. Of course, nothing is wrong with this; in fact, it might even be an important aspect of our survival instinct that is programmed into our genes. Forming a picture in our mind of our environment will guide our decisions and improve our chances of survival. However, we should always be keenly aware that our worldview today could be very wrong and that we might have to revise it in the future.

Presently there are a few clear signs that this revision is imminent. It is possible that we during the twentieth century moved away from a reasonable worldview. Since there is no explanation for the progression of time, one concludes that our senses must somehow deceive us. Yet the reason why we cannot explain it might simply be that something fundamentally important is missing in our understanding.

This book confronts these issues head-on by proposing a new and simpler, yet elegant, worldview. I will show that a new cosmological model exists that resolves numerous issues, including the progression of time. The universe might not be as mysterious and complicated as is currently believed; it could be both logically consistent and beautifully simple, although perhaps totally unexpected. This new cosmos model implies a new kind of process that influences the *scale* of space and time and explains the connection between General Relativity and Quantum Theory as well as the origin of the inertial force.

Quoting Carl Sagan from his book *Cosmos* (1980, p. 333):

Science...has two rules.

First: There are no sacred truths; all assumptions must be critically examined; arguments from authority are worthless.

Second: Whatever is inconsistent with facts must be discarded or revised. We must understand cosmos as it is and not confuse how it is with how we wish it to be. The obvious is sometimes false; the unexpected is sometimes true.

The new scale expanding cosmos model yields a simpler, more coherent and self-consistent account of the cosmos that better agrees with astronomical observations. Therefore, it isn't just an alternate theory; it is a better theory.

This book will undoubtedly be met with considerable skepticism by specialists in the fields of cosmology and physics because it challenges well-established "truths." Some might even reject it out of hand since it threatens to pull the rug out from under established epistemology going all the way back to the time of Galileo. This is unfortunate, since it makes me a heretic, but I think you might find my ideas interesting and, if I am right, even important.

I was trained in physics and engineering and have arrived at the conclusions presented in this book not by speculation or wishful thinking but by careful, scientific, investigation of the implications cosmological scale equivalence would have. To my astonishment I gradually came to realize that current physics cannot handle a dynamic spacetime scale as it would be experienced by an observer participating in the scale-expansion like we all do. In other words, Western physics based on Newton's laws implicitly assume that the scale of existence always has remained the same. The possibility that this may not be the case could forever change our world-view.

This book is divided into eleven chapters:

Chapter 1: **Spacetime-Scale Equivalence** introduces the concept of cosmological scale-equivalence as a previously unexplored additional degree of freedom. It implies a new aspect of motion since objects may move not only in the four dimensions of space and time but also in scale, making the world five-dimensional.

Chapter 2: **Justifying the Scale Expanding Cosmos (SEC)** The concept of scale-equivalence in the context of cosmological expansion is explored by introducing the SEC theory together with related subjects. Some may object to my use of the word "theory," which usually is reserved for more established ideas. However, it is a theory in the same sense as the SCM is a theory. The summary of the SEC theory given in this chapter is intended for the reader who might not be interested in the observational and theoretical details justifying the SEC.

Chapter 3: **A Few Unfamiliar Aspects of the SEC** are introduced since the SEC theory implies new physics.

Chapter 4: **Problems with the SCM and Their Resolutions** highlights several problems with the SCM that are resolved by the SEC model.

Chapter 5: **Observational Evidence in the Solar System** presents evidence for the new model found in the solar system.

Chapter 6: **New Physics of the SEC Model** presents new aspects implied by the SEC that to some extent would revise known epistemology.

Chapter 7: **Motion and the Origin of Inertia** offers an explanation of what causes the inertial force and suggests that the theory of Special Relativity might be in need of revision.

Chapter 8: **Quantum Theory and Its Link to General Relativity** introduces a previously missing link between the theory of general relativity and quantum theory by showing how quantum theory may be derived from general relativity. It also gives a physical, ontological, interpretation to the quantum mechanical wave-functions.

Chapter 9: **The SEC in Relation to Current Physics** places the proposed theory in a historical perspective.

Chapter 10: **Bits and Pieces** presents a number of my personal comments and ruminations over the years.

Chapter 11: **A Missing Dimension** summarizes this book by suggesting a new worldview, which might forever change our perception of the world.

Throughout this book, I reference published papers, which may be found at the end of the book, and which may be consulted for further study. My main objective with the book is to stimulate interest in a new worldview. Perhaps it will motivate you to follow in my footsteps into the unexplored territory of the unknown.

In introducing an idea like cosmological scale-expansion I realize that it cannot gain scientific credibility unless it is modeled by known theory, which allows us to make testable predictions and to compare these predictions to our observations. Thus, the new theory must be based on known science. And, if the idea is truly new connections to the already known must be established. Without this background support the credibility of the new idea will suffer.

While reading this book, you will enter into a new world. There might be a previously unsuspected aspect of the cosmos that dramatically changes the perspective of our existence. We participate together with all other existences throughout the cosmos in the dynamic process of becoming. We exist in a world where all is one and the progression of time is the engine of all existence.

Cosmological Expansion and Scale-Equivalence

IN QUESTIONS OF SCIENCE, THE AUTHORITY OF A THOUSAND IS NOT WORTH THE HUMBLE REASONING OF A SINGLE INDIVIDUAL.

GALILEO GALILEI

How We Got on the Wrong Track

I assume that you are familiar with the main justification for the Big Bang idea, which is the cosmological redshift (see sidebar).

The frequency of light from galaxies is diminished, or "redshifted", roughly in proportion to the galaxies' distances from us, which some have interpreted as a Doppler-type effect caused by the motion of galaxies away from us. With this assumption, the Big Bang idea was born. If galaxies are moving away from each other, and have always been moving away, they must have been very much closer together in the past. Extrapolating this even further backward in time, we end up in an infinitely dense state—the SCM creation event, that is, the Big Bang.

The Redshift

Light received from distant sources in the universe (like a galaxy) is "redshifted" in the sense that its spectrum is shifted to lower frequencies. The amount of redshift is measured by observing the frequency of a spectral line in relation to its nominal frequency measured in the laboratory. The formula for redshift is:

Redshift = [(nominal frequency)/(observed frequency)] − 1.

The redshift is close to zero for nearby sources and becomes positive for distant sources.

This model of the universe also gained support from the fact that a spatially expanding universe may be modelled by General Relativity (GR), which gave the model physical credibility. Two additional observational findings seemingly supported the Big Bang scenario: the light element abundances and the cosmic microwave background (CMB) radiation. Advocates for the Big Bang theory estimated the proportions of light elements like hydrogen, helium, and lithium that could have been created in the Big Bang event and found that they seem to agree fairly well with what actually is observed in our present universe. According to the SCM the CMB radiation was emitted immediately after the Big Bang creation event and is now reaching us from incredibly vast distances.

The cosmological redshift, the light element abundances and the CMB are the three pillars upon which the SCM rests.

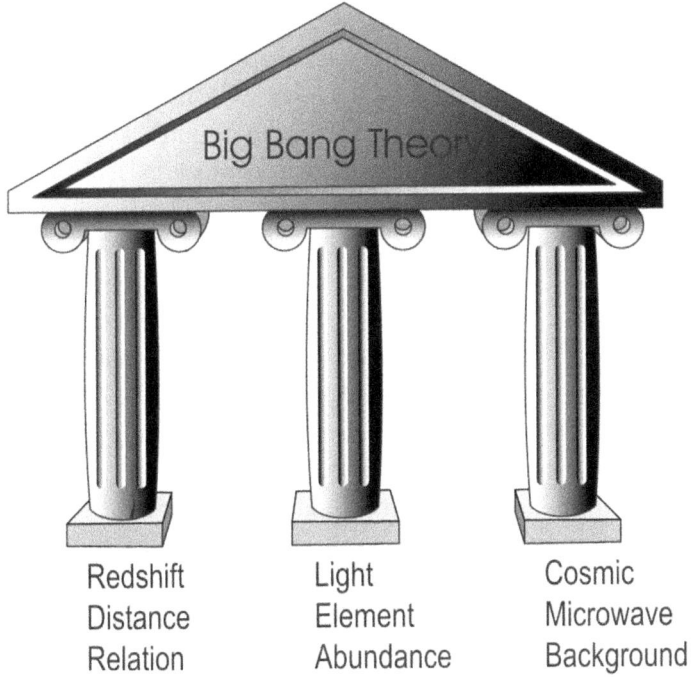

Figure 1: The three pillars of the SCM

Regarding the CMB, which was believed to be remaining radiation from the Big Bang, its temperature was initially estimated to be about 50 kelvin. When the CMB temperature was eventually found to be about 3 kelvin instead of 50 degrees, the Big Bang scenario was adjusted to accommodate this new finding.

In this context, it should be noted that prior to discovering the CMB radiation, several independent physicists had estimated a cosmological temperature close to the actual CMB's 2.7 kelvin, assuming temperature equilibrium in a static universe. The first calculation of the temperature of space, based on the energy of starlight, was done by Sir Arthur Eddington in 1926. He arrived at 3.2 K. A few years later Ernst Regener studied the ionization caused by the cosmic radiation and concluded that intergalactic space, where starlight is negligible, was filled with a background of temperature 2.8 K.

The adjustment of the SCM temperature to fit the observational data set the stage for a development that still continues: whenever observations disagree with some particular prediction of the SCM model, the model is adjusted to patch up the disagreement. What cannot easily be accommodated is often blamed on "evolution," with the explanation that the reason observations from the faraway earlier universe do not agree with the SCM model is that the universe was different in the past. If this does not work, discrepancies are "resolved" by additional speculative features such as mysterious dark energy, accelerating expansion, and so on.

It is not unusual that a "patch" that attempts to resolve one particular problem with the SCM conflicts with another patch attempting to resolve a different problem. Of course, this is not science, but since nobody can ever know anything about the conditions during the Big Bang creation event, it is always possible to explain away any discrepancy by adjusting the model or by invoking evolution.

During the early 1900s, a different model gained its own followers, mainly in England; this was the so-called

Steady State theory. The Steady State supporters accepted the expanding space idea, but instead of the Big Bang creation they proposed that new matter is continually being created in order to fill the expanding voids between galaxies. They argued that a steady, slow creation of new matter, particle by particle, isn't as strange as the instantaneous creation of all matter in the universe in one single event. By the Steady State theory, the universe could keep expanding forever without the Big Bang. In the 1950s, the debate ran hot between these two camps, but the SCM eventually won out because of the CMB.

If the CMB is the afterglow of the creation event, it should have certain characteristics; it should have a Planck black-body spectrum, the same type of electromagnetic spectrum that one finds in a dark cavity here on Earth. But this particular spectrum is very difficult to explain in the Steady State scenario. When measurements eventually showed that the CMB spectrum actually is very close to the black-body spectrum, the Steady State theory lost ground and the SCM theory became the accepted cosmological paradigm. This happened some 60 years ago.

However, the more we learn about the universe from new observations, the clearer it has become that the SCM model simply does not agree with the cosmos we see. Science is now confronted with a very serious problem. An open letter to the science community challenging the SCM was published in May 2004 in New Scientist, signed by thirty-three well-known researchers. This letter now has over five hundred underwriters [http://cosmologystatement.org/].

This book will explain where and how we went wrong with the SCM.

A New Degree of Freedom

Since the dawn of human civilization, we have been unaware of a most fundamental aspect of our existence, an aspect of the world that when it becomes fully known and accepted may lift a veil that has hidden its true nature. It answers deeply philosophical questions regarding the origin of the universe and the mysterious progression of time.

Truths hidden in plain view often are the most difficult to acknowledge because they hide under a blanket of the preconceived; they are not even recognized or questioned because they are aspects of our existence we take for granted. There are many examples of this from the past; for example, the Ptolemaic worldview never questioned that the Earth was fixed and at the center of the universe, the American continent simply did not exist in the minds of Europeans before its discovery. There are unknown questions that never are posed.

We live our lives with ideas and conceptions based on what we know. This is natural; how can we ever take into account what we don't know? However, it is likely that the universe cannot be explained by what we know. This has been true in the past and is still true today.

Taking stock of our predicament as human beings, we should, and must, acknowledge that we really don't know very much yet; we will come to know much more in the future. Accepting this fact is uncomfortable for us because it implies that we are actually basing our opinions and actions on ideas that likely are only partly right and even may be outright wrong.

The time may have come when the rug will be pulled out from under what we currently, and proudly, believe to be irrefutable scientific knowledge. The fortress of entrenched science could crumble with the collapse of one of the pillars at its foundation.

This pillar is the unchallenged belief that the scale of space and time always has remained the same.

In the past it has never occurred to us that the scale of the universe might change with time. I am not aware of anyone who has challenged this ingrained belief. The giants of science have never even given it a second thought. To my knowledge the possibility that the world might be expanding in a scale-dimension beyond the four dimensions of space and time has never even been considered. Yet it might be the most fundamental aspect of all existence.

Spatial and temporal increments might expand with time.

Since scientific epistemology overlooks this new degree of freedom, fundamental laws at the core of science that we believe to be absolutely true, like for example Newton's first law of motion, may actually be false. The same goes for the laws of thermodynamics because the universe could be in perpetual motion. How I arrived at these heretical conclusions is the subject of this book.

Let me take you back to December 1992, when, at the airport in Seattle, I found a little book with the title *Ancient Light* written by the astrophysicist Alan Lightman. I learned to my astonishment that all is not well with the SCM based on the Big Bang creation idea. In the evening after arriving in New York, lying in my bed at the Marriott Hotel on Times Square, I suddenly got a wonderfully simple idea.

It was this:

If not only space, but also time, were to expand in the cosmological expansion, it would not change distances measured by timing rays of light because increasing distances would then be compensated for by a slowing pace of time. Since the duration of a second increases when the scale expands and distances increase, the time it takes for light to cover a certain distance would remain the same.

This would mean that the cosmos could expand without really expanding!

At first, I did not realize that the expansion of space and time also must imply the expansion of all material objects including our own bodies. However, I found that expansion of spacetime also implies expansion of matter.

This may be realized by noting that changing the scale by a constant scale factor will not change Einstein's field equations of GR; they remain identical, with the implication that all laws of physics remain the same. We might say that GR is "blind" to discrete scale changes. This is not surprising because GR is all about geometry, and geometry does not distinguish between scales: a sphere is a sphere, regardless of its scale.

Another way to see this is to consider a location just below the surface of the Earth. At this location, the mass density is obviously quite high. However, when spacetime expands, a coordinate location moving together with the expanding spacetime grid would, if the Earth didn't expand, eventually move above the surface of the Earth where the mass density is much lower. This disagrees with GR, by which the mass density always should remain the same. We conclude that the Earth must expand together with spacetime. Therefore, all material objects must expand together with spacetime.

The cosmos expands in scale while always remaining the same!

Albert Einstein

Albert Einstein (1879–1955) is probably the best known and most admired scientist of our time. He is the father of both the Special Relativity theory and the General Relativity theory. In 1905, he published three important papers: an interpretation of the photoelectric effect based on the hypothesis that energy of light comes in discrete quanta; the statistical theory of Brownian motion; and the Special Relativity theory. He received the Nobel Prize for the first two papers in 1921. In 1915, he published the General Relativity theory in its final form.

Einstein's most unusual quality was his strong conviction that the secrets of Nature are accessible to the human intellect and may be revealed to a mind free of conventions and preconceptions. His approach was to ask himself what would be the simplest and most logical design of the universe.

This led him to the conviction that the force of gravitation and the force of inertia must be manifestations of the same phenomenon, which would explain why inertial mass and gravitational mass are equal. Gravitation was no longer a mysterious force reaching out over empty space but instead a feature of spacetime itself.

The General Relativity theory provides a strong connection between spacetime and matter and suggests that spacetime may contain energy. This book will further suggest that although Einstein made an invaluable contribution he may have overlooked a fundamentally important aspect of all existence.

This means that the expansion occurs in a dimension beyond space and time and could continue forever; the world could be in perpetual motion in scale in obvious violation of the laws of thermodynamics. The scale-expansion has not been acknowledged earlier because the scale changes very slowly. In relation to a hypothetical universe with constant scale it increases by 50 percent in about ten billion years. However, a co-expanding observer, like we all are, would not locally perceive this changing scale. Yet, as we shall see, its cosmological implications are profound and readily observable.

This turned out to be the first glimpses of a mountain of previously undiscovered knowledge, much of which will obsolete ancient misconceptions. I will reveal some of them beginning with cosmology, since finding an explanation to the universe has always been at the core of scientific exploration. However, as we shall see, there are further implications of fundamental importance.

Cosmological Scale-Equivalence

The most significant aspect of the SCM is that "metrics" of space expand while the temporal "metric" always remains the same; in other words, the SCM assumes that time has always progressed at the same pace. As used in GR a "metric" defines the scale of one the four dimensions of spacetime. Expanding spatial metrics would mean that the spatial scale was smaller relative to the temporal scale in the past so that space appeared to be compacted in relation to our current spatial scale. Extrapolating this backward in time leads to the Big Bang scenario.

However, we have seen that there could be another expansion mode; in addition to the spatial scale, the temporal scale could also expand in uniform. This is the SEC model. It gives a different interpretation of the cosmological redshift. Rather than resulting from galaxies moving away from us, the redshift is now caused by gravitational-type action on photons reaching us from the distant past; more about this later. By the SEC theory the cosmos is four-dimensionally (4D) "scale-equivalent" so that it remains physically the same relative to its inhabitants regardless of the scale of space and time. We might say that the cosmological scale-expansion of both space and time preserves the relationship between space and time and according to GR, all physics. Consequently, to co-expanding inhabitants that we all are, the cosmos may always have been the same in relation to us.

The cosmos is dynamically scale-equivalent. This is illustrated in figure 2.

Figure 2: The expanding space and time

This scale-expansion model might seem quite natural since we now know that the cosmos is four-dimensional, but it has been overlooked in the past because it cannot be modeled by GR. We will return to this important fact later in the text.

Introducing the SEC Model

The main reason why challenges to the SCM have failed in the past is that no competing theory has been on the horizon after the demise of the Steady State theory. Even if we sense that the SCM must be wrong, challenging it in the absence of a better model is difficult. However, this has now changed; with the SEC, a better model now exists. This new model resolves many cosmological puzzles and is so simple and elegant that you might wonder why it hasn't been considered before. Here is the SEC in just a few words:

The universe expands by changing the scale of both space and time while conserving all physics via scale-equivalence.

This 4D-expansion model is illustrated in figure 3, where the third spatial dimension is suppressed.

When the length of a meter (or foot) expands, the pace of time slows down, making time intervals like a second longer in proportion to the length.

This new model explains all cosmological observations, including the Cosmological Microwave Background (CMB) radiation, without resorting to speculation or evolution. In the SEC, the CMB is simply equalized electromagnetic radiation (thermalized radiation) including starlight, which over eons has assumed a black-body spectrum by redshifting (see sidebar). Four-dimensional scale-expansion, which preserves the black-body spectrum, makes this possible (see [Masreliez, 2004a] for more information).

SCM model; space expands but not time

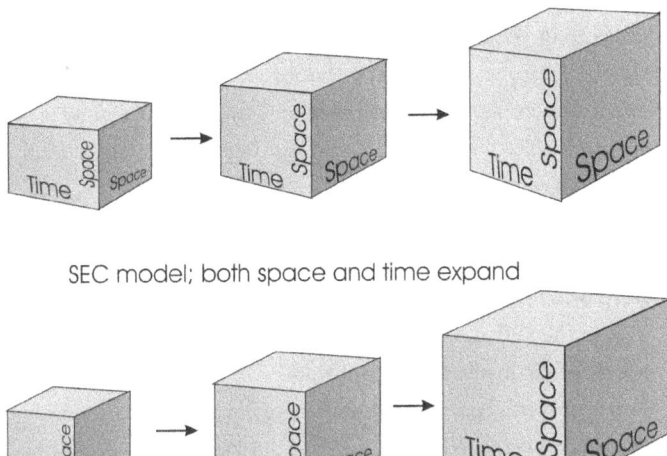

SEC model; both space and time expand

Figure 3: The SCM and SEC model scale expansions

The Cosmic Microwave Background

Around 1960, cosmologists Robert Dicke and P. J. E. Peeples at Princeton University speculated that if there were a Big Bang, the radiation from the very hot universe immediately afterward might still be detectable, reaching us from extreme distances. They thought that this electromagnetic radiation ought to have a black-body spectrum with a temperature of around 10 kelvin. They started to develop a special radio receiver capable of detecting cosmological radiation at microwave frequencies.

Around 1963, two scientists at Bell Labs, Arno Penzias and Robert Wilson, had built a receiver to be used for radio astronomy. However, they could not make it work properly due to an unexplainable static noise. They tried to find the source of this noise for almost a year—this went so far as cleaning pigeon droppings from the receiver antenna. When they complained about their problem to a friend they learned of the work at Princeton and realized that the strange radio noise might be of cosmological origin. Their subsequent investigation led to their discovery of the *cosmic microwave background* (CMB).

As a result of their discovery, the two Bell Lab scientists were awarded the Nobel Prize (while ironically no recognition was given to the people at Princeton).

In the SEC theory CMB radiation results from electromagnetic energy in thermal equilibrium. All the radiating sources in the universe continuously add energy, which is continuously dissipated via the redshift. These two mechanisms reach equilibrium at the CMB temperature of 2.73 kelvin.

The light element abundances may be explained as resulting from active galaxy nuclei and quasars, which often are seen ejecting gas in jets (see discussion in [Masreliez, 2004b]). Furthermore, the SEC theory tells us that a dynamic spacetime scale is causing the progression of time and provides the missing connection between general relativity and quantum mechanics [Masreliez, 2005a]. And it turns out that black holes cannot form [Masreliez, 2004c]. Although we have observed motions of stars close to the cores of galaxies that suggest very large mass concentrations, a "naked" black hole has never been seen. As we shall see, the ejection of gas from active galactic nuclei suggests a different explanation.

The dynamic scale of the SEC theory also suggests that the scale of spacetime might change with motion in space. If this were true, the inertial force we feel when accelerating may be explained as being a curved spacetime phenomenon akin to gravitation. This line of reasoning is developed in Chapter 7.

Often, hidden preconceptions cause us to misunderstand the true nature of the world. Einstein showed that the pace of time might differ in a relative sense for objects in motion and that gravitational fields also might influence the pace of time. Thus, he challenged the presumption that the metrics of space and time are, and always have been, the same everywhere. The SEC theory goes one step further by proposing that the scale of time also might change *dynamically with time*. When space expands, the pace of time slows down. One reason why this possibility might have been overlooked in the past is that GR cannot model a decreasing pace of time; it assumes that the pace of time always remains the same for an object at rest.

CHAPTER 2

Justifying the Scale Expanding Cosmos

This chapter goes deeper into the new theory, showing how a number of previously irresolvable issues in physics and cosmology may be resolved by the SEC model.

What is your opinion about modern cosmology? Maybe you have seen speculations that our universe may have been "spawned by" a mother universe or that it was created in an extremely huge quantum fluctuation. The problem with this, as with any other way of trying to explain the creation of the universe, is that we never will be able to confirm it; the creation event will always remain enigmatic.

Before the Big Bang idea gained popularity in the middle of the twentieth century, most people believed that the world exists eternally. In fact, an eternally existing universe was championed by Parmenides some 2,500 years ago. He argued that either the universe exists or it doesn't exist. If it exists, it obviously cannot have been created from something that does *not* exist, because nonexistence means nothingness.

He therefore logically concluded that the universe must always have existed.

This line of reasoning makes perfect sense, but the mind reels when confronted with the idea of eternal existence. Yet, the SEC theory allows perpetual existence. This new model is mathematically simple, is internally self-consistent, and offers many advantages over the SCM theory; for example:

- The SEC is a comprehensive theory based on perpetual cosmological scale-expansion.

- The SEC better agrees with actual astronomical observations than does the SCM model.

- The SEC is internally self-consistent. It addresses a range of existing problems that previously appeared to be unrelated.

- The SEC explains what makes time progress.

- The SEC provides the missing link between GR and QT and explains why there is a quantum world.

- A new dynamic process related to the SEC model explains what causes the inertial force.

This new theory gives a very different, yet simpler, worldview. It introduces new ideas, which are supported by observational data and conceptual clarity. You will find that the SEC theory is elegant compared to the SCM patchwork.

The Age of the Universe

According to the SCM model, the universe appears to have begun with the Big Bang about 14 billion years ago. However, by the SEC model, the duration of a year was shorter in the past. Therefore, the age of the universe may be viewed as the sum of an infinite number of years going backward in time, each preceding year being slightly shorter than the following.

It is well known that the sum of a geometric series with quotient less than one is finite even if the number of terms is infinite. Therefore, an infinite number of years into the past may add up to a finite number of years with the current length of year, as illustrated in figure 4.

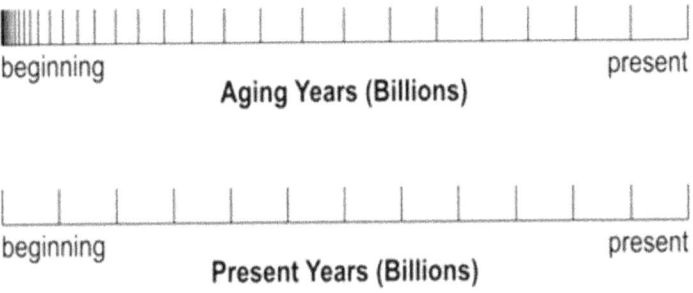

beginning   **Aging Years (Billions)**   present

beginning   **Present Years (Billions)**   present

Figure 4: The SEC and SCM time scales
The upper part of the figure shows the progression of time in the SEC and the lower in the SCM.

The SCM assumes that the duration of a year has always been the same as the current duration, which gives the impression that the universe began some 10-14 billion years ago. This "age of the universe" is called the Hubble time, and for simplicity I will in the following use the value 14 billion years since the currently accepted estimate is 13.8 billion years. However, in the SEC the Hubble time is a cosmological constant that has nothing to do with the age of the universe. In the SEC model the years get shorter and shorter going back in time relative to the current year. An observer living in the SEC 14 billion years ago would also have concluded that the universe is 14 billion years old based on the length of the year 14 billion years ago. This is reminiscent of someone onboard a ship on the ocean who sees the horizon at the same distance all the time. The ocean's horizon is due to the curvature of the Earth and the horizon of the observable universe is caused by the curvature of spacetime in the SEC. The horizon of the universe is always 14 billion light-years away regardless of spatial location and epoch. However, the size of the universe may be unlimited both in space and time.

Justifying the SEC Model

It has been said that extraordinary claims demand extraordinary proof. Let's recall the following thought experiment. Imagine that you are an intelligence facing the task of creating something—for example, an apple in a void of nothingness. What size should you make this primordial apple? It occurs to you that since there is no reference at all in nothingness, you could actually make it any size, like a pea, a basketball, or even like the Earth, provided you create all atoms in the apple to scale. And, since atoms oscillate at specific frequencies, you could also adjust the pace of time to match the spatial scale. Beginning with this primordial apple, the rest of the world could then be created in proportion. If this isn't true, something in nothingness must determine the scale of things, which contradicts the concept of nothingness. We must conclude that no particular cosmological scale is preferred and that physically equivalent universes might exist at different scales.

This reasoning suggests that the cosmological scale of space and time might be changing with the progression of time. This is the essence of the SEC theory.

A scale-expanding cosmos has no beginning or end; the scale may continually expand forever, and as experienced by a co-expanding observer, all epochs would appear physically equivalent. This conclusion is also supported by GR since the field equations are identical for different scales. Thus, it is quite possible that there was no Big Bang creation event, and that the universe has always existed, eliminating the most troublesome aspect of the SCM theory.

An observer in the SEC, like you and me, expands together with the universe and will not notice the expansion locally, since everything else in our environment expands at the same pace, including material objects and the duration of a second. However, although we cannot notice the expansion locally, its effects can be seen in the redshifted light from faraway galaxies. As already mentioned, this redshift is not due to recession; if we were able to extend a very (very!) long measuring tape between two galaxies, we would find that they remain in the same relative positions since the tape expands along with everything else. And if we measure the distance by timing a light photon on its way between galaxies, we would measure the same constant time interval, since the pace of time slows down when space expands. Thus, relative positions of galaxies would remain the same on the average; you might say that the universe expands without expanding! The cosmological explanation mode is not in space but in *scale*, which does not change the 4D relationship between space and time; we might say that the expansion occurs "beyond space and time."

Furthermore, since universes of different scales are physically equivalent, the scale-expansion may proceed without cosmological aging. Perhaps you might object to this because it seems to violate the laws of thermodynamics by suggesting that the universe is a perpetual motion "machine." However, this conundrum is resolved by the expanding temporal scale, *which has the effect of inducing cosmological energy*. I will return to this important but unfamiliar aspect of the SEC model shortly.

Although the idea of cosmological scale-expansion seems natural to most people without scientific training, people in science may have a hard time accepting this simple idea, because *cosmological scale-expansion, which makes all epochs equivalent, cannot be modeled by GR.* Since GR and QT are the two central pillars of modern physics, considering anything that might violate GR is unpalatable.

But scale-expansion is such a simple and natural idea that we should not abandon it simply because it cannot be modeled by GR. The SEC offers better interpretations to our astronomical observations which should make us question the SCM and GR.

Remember what Carl Sagan said in his book *Cosmos*:

"We must understand that the cosmos is what it is and not confuse how it is with what we wish it to be. The obvious is sometimes false; the unexpected sometimes true." (p. 333)

There is an interesting parallel from the past. Orbits of the planets were once believed to be circular or, if not circular, to consist of circles within, or upon, circles—the so-called epicycles. This was based on the ancient belief attributed to Aristotle that circular motion is the most perfect kind of motion and that, therefore, the planetary motions should be described by circles. We might say that the postulate of circular motion was then the only acceptable theory of the cosmos. Then Johannes Kepler came along and violated this constraint by suggesting ugly, elliptical, orbits based on observations.

Today, GR is the generally accepted theory and consequently we believe that the cosmos should be modelled by GR, which leads to the SCM. However, history might repeat itself; we might find that GR is *not* the final word. Like the circles of Aristotle were replaced by Kepler's ellipses, GR might be replaced by an improved theory from which the SEC model naturally follows together with explanation to numerous cosmological puzzles.

The SEC Implies New Physics

Let's try to model four-dimensional scale-expansion. We have to come up with a way to describe by mathematics how the duration of a second changes with time, but we wonder how may time possibly change relative to itself? However, if we instead of time use *the cosmological scale as an additional parameter*, we could model a scale that increases with time in GR, which, in effect, indirectly would increase the duration of time intervals via the scale. The problem is that this kind of dynamic process is new to physics because the motion is not in space but in scale.

In GR, the physics of space and time is believed to be independent of our choice of coordinates; and if we change the scale of spacetime, the new expanding temporal scale may by GR be transformed into new situation with a temporal scale that does not change with time. As a result we end up with a purely spatial expansion like in the SCM, and lose the fundamental feature of cosmological scale-equivalence. Therefore, GR cannot model a scale expanding cosmos *as experienced by an inhabitant*. The freedom of changing the coordinates is a particular aspect of GR called *covariance*. I will return to this later in the text.

Covariance implies that GR does not distinguish between a cosmos with a beginning of time and cosmos of eternal existence. Obviously this cannot be right. It seems self-evident that something must be wrong here; the emperor has no clothes. We now know that the scale of time is not arbitrary since atomic oscillations define atomic time, which nowadays is the temporal standard used in science. Consequently, Nature defines its own temporal standard that cannot be changed by coordinate transformation. And, as we shall see, atomic time implies eternal existence.

The problem of how to model cosmological scale-expansion by known physics occupied my thoughts for about two years. I found that this had also been an insurmountable obstacle for other investigators, who in the past had considered scale-expansion. Finally I concluded that *GR must be incomplete* since it cannot model scale-equivalent cosmological expansion *as perceived by a co-expanding observer*. This failure is obviously a real problem if we all are co-expanding observers because, in this case, relying on GR will not work!

At first, this was a disappointment, but I thought that scale-expansion was such a simple and pure idea that this expansion mode should be possible, even if it cannot be modeled by GR. After further investigating the properties of the SEC, and finding that this new model accurately describes a number of astronomical observations far better than GR can do, I gradually became convinced that GR should be generalized to include *discrete* scale adjustments. I found that if the cosmological scale were to change via continuous scale-expansion in space and time combined with incremental scale adjustments, the SEC could be modeled by GR! Later I found that the SEC also may be modelled by adding a new "dimension" in the form of a dynamic spacetime scale to the four ordinary dimensions of space (three dimensions) and time (one dimension), forming a new five-dimensional cosmos model. I will call this 5D space *"hyperspace"* to distinguish it from 4D spacetime.

GR is "blind" to discrete, stepwise scale changes because Einstein's field equations remain unaffected. Therefore, if the cosmological expansion were to include discrete scale increments, treating the scale as a fifth dimension beyond the four spacetime dimensions, GR could still be used. If we accept this solution to our modeling problem, we will also gain three very important advantages.

First, the progression of time may now be explained: *the incrementally increasing scale could cause time to progress*. This would also explain why it in the past has been impossible to model the progression of time by GR. GR does not "see" the expanding scale; it takes place beyond the four dimensions of spacetime.

The second advantage is that a dynamically expanding spacetime scale would provide an explanation of the inertial force we feel during acceleration (Chapter 7).

The third advantage is that it may offer the missing link between general relativity and quantum theory (Chapter 8)!

Cosmic Drag

One direct consequence of the SEC model is that relative velocities of freely moving particles (I take *particle* to mean any object with positive rest mass) will diminish over time; relative velocities well below the speed of light will decrease exponentially with a time constant that equals the Hubble time. The Hubble time is the age of the universe in the SCM model, which is about 14 billion years. Relative velocities will diminish by 50% in about ten billion years. It seems like a particle is subjected to a "drag force" when it actually is freely moving.

Cosmic drag may be seen as being gravitational in nature; as if moving out of a weak gravitational field resulting from the scale-expansion.

For velocities well below the speed of light the Cosmic Drag relation is:

$$v = v_0 \cdot e^{-t/T}$$

Here, v_0 is the initial velocity, t is atomic time, and T, the Hubble time. This relation is derived in Appendix I.

Cosmic drag is a new phenomenon unique to the SEC model that may be visualized by a thought experiment: Considering an ant crawling over a slowly expanding coordinate grid on a surface. Although the ant keeps moving at the same pace it will lose grounds as measured with the expanding coordinates. It seems that the ant is slowing down when in reality it is the measurement scale that is increasing. Its speed diminishes in proportion to exp(-t/T) due to the expanding space. But, how does the temporal coordinate enter into this picture? It seems that the slowing pace of time would cause the measured speed to increase and compensate for the spatial expansion. However, although each little step by the ant remains the same, the pace of these steps slows down by the factor exp(-t/T) due to the expanding temporal scale, which would cancel out the temporal expansion effect. As a result the ant's speed slows down in proportion to exp(-t/T), which is the cosmic drag effect.

Cosmic drag also acts on photons and explains the redshift as being a gravitational phenomenon induced by the scale-expansion. Since the scale of spacetime was smaller in the past, photons arriving from the past have been moving against a gravitational field induced by the cosmological expansion. Therefore, a weak acceleration acts backward in time and space, diminishing photon energies. This cosmic drag diminishes relative motion and causes particle velocities to slowly decline. It also means that the relative velocities between galaxies tend to be quite small. As estimated from observations, they typically are less than one percent of the speed of light, an observational fact that has been difficult to explain in the SCM scenario where numerical simulations indicate that relative velocities ought to be much higher than what is observed.

Cosmic drag will also slow down rotating motion, causing angular momentums to decrease exponentially with time. As a consequence, stars in a spiral galaxy will follow shallow spiral trajectories on their inward paths, and the gravitational attraction between them will form the beautiful spiral arms we observe. Thus, the SEC theory would also explain the formation of spiral galaxies, which has been a previously unresolved problem (see further below and [Masreliez, 2004b]).

Cosmic drag should also influence the planetary motions in our solar system, causing the planets to slowly approach the Sun in shallow spiral orbits. According to the SEC theory, the Earth currently approaches the Sun by about 10 meters per year, and the angular velocity accelerates by about 1 arc-second per century squared ($1"/cy^2$). This effect is extremely small and has therefore not been noticed until very recently, after Atomic Time (AT) became available in 1955.

This spiraling motion of the planets has not been detected in the past because the concept of time in astronomy was determined by the rotation of the Earth (Universal Time) and by the motion of the Earth around the Sun (Ephemeris Time). Obviously, if one defines the length of the year as the time it takes for the Earth to circumnavigate the Sun, any possible acceleration of the Earth in its orbit around the Sun will become undetectable simply because of the way we have defined time. In other words, it will always take exactly one year for the Earth to circle the Sun because this is how we have defined our temporal standard! This approach made it impossible to detect the spiralling motions in the past before introducing AT into astronomy in 1955. But now, after making observations for more than fifty years with access to AT, observational discrepancies are starting to appear. This currently is a mysterious, unresolved problem in astronomy, which is a very interesting "breaking news" situation. It may soon confirm the SEC theory. See chapter 6 for more information about this development.

We could be facing a new "Copernican revolution" that will change our worldview forever.

The Question of a Cosmological Reference Frame

The reader might here have realized that the existence of cosmic drag would violate Newton's first law of motion, stating that freely moving objects will continue to move at the same velocity forever. This is true; cosmological scale-expansion of both time and space implies that Newton's first law no longer holds as stated by Newton, which also would mean that conservation of kinetic energy no longer holds true. However, in the SEC this classical conservation law is replaced by conservation of energy-momentum, which takes into account the changing spacetime scale. The loss of kinetic energy is then balanced by the energy added via the slowing progression of time so that the combined total energy remains the same. In other words, the conservation of energy and momentum still holds four-dimensionally.

The existence of cosmic drag would invalidate Newton's first law of motion, but on the flip side, it would resolve a festering, still unresolved problem since the days of Newton—the question of a cosmological reference frame.

In his famous spinning bucket experiment, Isaac Newton observed that the surface of the water in a spinning bucket becomes concave and concluded that the bucket somehow "senses" that it is spinning. But spinning relative to what? It is not the Earth because the planets are subjected to the same centrifugal force in their motion around the Sun. And now we know it is not the Sun, since stars in a galaxy are subjected to the same force. Newton concluded that a frame of absolute universal rest must exist, and this became the subject of a celebrated debate between Samuel Clarke in England, who spoke for Newton's position, and Gottfried Wilhelm Leibniz, who contended that all motion is relative and that there is no absolute cosmological rest frame.

From the time of Newton until Einstein's special relativity theory appeared in 1905, the general consensus was that a cosmological reference frame exists as defined by the "aether," which was believed to be some kind of undefined plenum in absolute rest carrying light and the electromagnetic field. Although Einstein at first did away with the aether in his SR theory, he only remained convinced of its nonexistence for a relatively short time—the ten years between 1905 and 1915. However, after introducing GR, he gradually changed his position. By the end of his life, Einstein thought that spacetime was a new form of aether that somehow serves as a reference for inertia. As we shall see, he may have been partly right.

Cosmic drag would resolve this problem by defining a cosmological spatial reference frame as the frame toward which all motion converges. Thus, the cosmological reference frame in the SEC is self-induced, caused by diminishing relative velocities and rotations. The scale-expansion creates a particular preferred, absolute, universal spatial reference frame just as Newton thought. The scale-expansion also models the progression of time, which directly is reflected by the pace of an atomic clock. Together with absolute space, this defines *absolute references for space and time,* with the speed of light giving the relation between them.

This should be good news, since physics sorely needs a temporal reference frame in particular in the context of non-local actions in quantum theory and as a common cosmological temporal reference; see further [Masreliez, 2006a]. Such a reference frame would also explain why CMB varies so that its frequency is slightly higher in a certain direction and lower in the opposite direction suggesting a Doppler shift that indicates that the solar system is in motion relative to the very distant universe at about 370 km/sec. This is commonly referred to as the "dipole" of the CMB.

With this development, Newton's absolute space (and time) may make a comeback after a hundred-year detour that began with SR. As mentioned, this will also resolve the closely related mystery of what might cause the inertial force: the scale-expansion is occurring at all locations throughout space and time, and it explains how a particle may "feel" that it is accelerating relative to the cosmological background. This dynamic background reference might be present everywhere in the form of an expanding scale, and could resolve a long-standing mystery that has been intensely debated ever since the dawn of Western science.

The Origin of Inertia

What is causing the inertial force has been a mystery since the time of Newton. By his second postulate (law) of motion, $F = am$, he acknowledged the fact that a force is needed to accelerate an object, but he did not explain why this should be the case. And nobody since then has been able to give a generally accepted explanation to what is causing the inertial force we feel when accelerating.

Newton thought that the inertial force must be related to the gravitational force because both are proportional to mass, which allows the inertial, centrifugal, force to be balanced by the gravitational force—for example, in planetary motion. In other words, he assumed that the inertial mass and the gravitational mass are one and the same.

Einstein later agreed when postulating that inertia and gravitation are caused by the same phenomenon. In his famous elevator thought experiment, he argued that an observer inside an accelerating box far out in empty space would experience an inertial force similar to the gravitational force. This became his guideline when developing GR. He realized that a freely falling object is in the same situation as a freely moving object in an inertial frame; in both cases, there are no forces. However, he never found what is causing this inertial force. As we shall see, ironically the reason for his failure might have been his own Special Relativity, which in effect may have blocked the discovery of the origin of inertia.

The problem is that SR may have overlooked an important property of motion, which has not been recognized in the past: *a dynamic 4D scale*. If the spacetime scale were to contract in a relative sense during motion, the inertial force may be explained as a curved spacetime phenomenon similar to gravitation. Furthermore, this would also allow clocks to always run at the same pace in inertial frames. This is further discussed in chapter 7.

So, if spacetime is curved by acceleration, it *implies* Newton's second law of motion, which may then be derived from GR. Important features of SR would still remain intact and inertial frames are still physically equivalent.

If the scale is dynamic, the observational finding that time seems to run slower for moving objects *could be a relative phenomenon* caused by relative scale contraction. This would also resolve the Twin Paradox (see chapter 7) by allowing the twins to always age at the same pace. It would introduce a new important aspect of motion: motion would not merely change the location in time and space but also the spacetime scale.

Many have been keenly aware that something must be wrong with SR, but exactly what it is has escaped us in the past. The problem with SR is that Einstein, like everyone else before and after him, assumed that the relationship between moving coordinate frames may be modeled by coordinate transformation, which relates locations in two different frames in one-to-one correspondence. However, if motion influences the metrics, this aspect must also be taken into account.

The English professor Herbert Dingle realized that something is not right with SR as exemplified by the Twin Paradox and repeatedly tried to make people in science acknowledge this problem but to no avail. SR was at the time well established and had become gospel not to be challenged. Dingle was right when he said that ignoring the inconsistencies with SR would have very detrimental consequences for science. The confusion caused by SR with its rejection of absolute time and space may have hampered the development of physics and prevented the discovery of the origin of inertia.

The SEC Explains the Quantum World

The incompatibility of GR and QT is one of the most vexing problems of contemporary science. These two theories successfully model different aspects of the world, but they are starkly different, both in philosophical outlook and in scope. GR applies to gravitation and to cosmology while QT primarily deals with the sub-microscopic world. And GR is based on differential geometry, while QT uses different mathematics like quantum mechanics. Although these two theories describe different aspects of the same universe, it is perplexing that they are so different and that they are incompatible. The SEC theory provides a simple resolution to this puzzling dilemma, since it allows Quantum Mechanics (QM) to be derived from GR. QM as used here refers to classical QT that primarily is based on the Schrödinger equation. This development is presented in chapter 8.

The key to understanding the link between GR and QM is the incremental scale-expansion of the SEC theory. Scale-expansion means that the length of a fixed distance, such as a meter, slowly expands and that the pace of time slowly decreases. If this were a continuous process, we wouldn't notice it locally, but according to the SEC theory, it is an incremental process. The scale might expand a little and then "jump into" a new, slightly larger scale by a tiny, discrete step. This progression could take place at an extremely high frequency that cannot be measured by present technology.

Like a child who repeatedly outgrows her clothes and gets new, larger clothes, we may repeatedly grow out of our cosmological scale before jumping into a larger scale by abruptly changing the pace of time. This expansion process is of course a new concept and at first we might not believe it can be true. However, it is possible to model part of this process in GR by considering tiny, very rapid oscillations superposed on the scale (metrics). *We find that GR with oscillating metrics may describe the quantum world, see {Masreliez (2005a}.* Now we realize that the quantum world might be a direct consequence of incremental cosmological scale-expansion! This is further discussed in chapter 8 and in Appendix IV.

A New Look at Motion

Perhaps it is not difficult to understand why people over the ages have found it so hard to understand the nature of "motion"; in fact, motion has been an enigma ever since the very dawn of science in the ancient Greece. Motion means that an object changes location in space with time, but how does it do this? Does it sort of crawl from one location to another like an inchworm, or does it jump forward in tiny increments? This question bothered the ancient Greeks as exemplified by Zeno's paradoxes—for example, his Arrow Paradox, which Aristotle summarized:

1. When the arrow is in a place just its own size, it's at rest.
2. At every moment of its flight, the arrow is in a place just its own size.
3. Therefore, at every moment of its flight, the arrow is at rest.

This puzzle was believed to be "solved" with the development of differential calculus in the 1600s by which jump-increments could become arbitrary small, but we now know that this approach fails due to quantum theoretical constraints. It is eerie how Aristotle could so clearly have expressed a long forgotten key aspect of motion by realizing that the arrow actually is at rest in its local spacetime geometry, even during motion. This will be further discussed in chapter 7.

We now know that making the increments of motion arbitrarily small really cannot explain motion either, because QT tells us that there is a limit as to how small one can make these increments. We are still facing the fact that a rigid object seemingly has to jump in small increments in order to be able to move. This conceptual problem with motion has been "swept under the rug" and is currently being ignored by science because no definite resolution has been found. And, perhaps more significantly, treating motion by continuous differential mathematics has become so well accepted that it no longer is questioned. I fear that this is a mistake.

In this book, I suggest that motion influences the spacetime metrics; we might say that the geometry of material objects "morphs" during acceleration, which would explain the inertial force. However, this explanation is new and controversial; it would mean that methods in use today for mathematically modeling motion could be inadequate because they do not take into account scale changes.

A possible, but somewhat speculative, explanation to the mystery of motion is presented in Appendix VI.

C. JOHAN MASRELIEZ

Can Any Cosmological Model Really Describe the Universe?

A cosmological model expresses our desire to make sense of the universe. It attempts to describe properties of the universe based on what we know. In this, we are constrained by our presumptions, our biases, and our current level of insight. This has always been, and will always be, true; ancient peoples perceived the universe as filled with mythological creatures. We know more now, but we still must use familiar concepts to describe the cosmos. And we still invent mythological features to fill in gaps where our knowledge falls short, for example dark energy.

Science tries to explain Nature by constructing models from which various features maybe predicted. But these models can be no better than the "material" we use to build them with, this material being known and accepted epistemology.

Judging from our past, we know that older models of the universe were inadequate simply because people did not know enough, and we should realize that this still applies. Any model we can conceive of is bound to be incomplete or perhaps even wrong; we simply do not know enough yet, and we will never know everything.

The best we can hope for is a model that makes sense to us with our current level of understanding. A good model should agree with observations and be internally consistent. Any cosmos model that does better in this regard is an improved model even if it means that we have to use new building materials. This is how progress is made. With our current knowledge, the model should seem right, at least until we learn more. Therefore, we should not believe that any cosmological model is the last word. We should realize that a cosmos model is only a tool for us to try to make better sense of the world. The SCM and the SEC theories should be viewed with this in mind.

To you, the reader of this book, it will become obvious that the SEC theory is a better model of the universe than the SCM. The SCM does not even satisfy the minimal requirement that a model should agree with observations and be internally consistent. On the other hand the SEC theory agrees with all observations and explains many previously unresolved cosmological puzzles. It does this without resorting to strange assumptions like the existence of dark matter and dark energy, and observations agree with the SEC model's prediction without speculating on decelerating expansion or accelerating expansion, or on evolution.

Furthermore, it suggests a new type of process involving dynamic spacetime metrics, which not only models the cosmological expansion but also explains what causes the progression of time and the quantum world. And it suggests that the origin of the inertial force is to be found in dynamic spacetime metrics.

CHAPTER 3

A Few Unfamiliar Aspects of the Scale Expanding Cosmos

So many centuries after the Creation, it is unlikely that anyone could find hitherto unknown lands of any value.

—Spanish Royal Commission, rejecting Christopher Columbus's proposal to sail west

In this chapter, I discuss the details of the SEC model, with the hope of convincing the reader of its viability. Some passages of the material are a bit heavy, and I apologize for some overlapping with the material presented above, which here is restated for the convenience of the reader.

A New Interpretation of the Cosmological Redshift

As we saw, the currently popular cosmological model was motivated by the discovery in the beginning of the twentieth century that light from distant galaxies is redshifted roughly in proportion to their distances from Earth. The usual explanation is that the redshift is caused by the expanding spatial scale in the SCM that makes galaxies recede. It is generally believed that the redshift is a Doppler-type effect due to the outward motion of distant sources, which lowers the frequency of light toward the red portion of the spectrum. However, since by the SCM space expands while the pace of time remains the same it would take longer and longer for light to cover a certain distance. In other words, the different scale of space and time also implies that the speed of light should change during the cosmological expansion, which would contradict Einstein's contention that the speed of light remains constant in the absence of gravitational fields.

Einstein objected to the SCM's explanation to the redshift. Here are his comments taken from his book *The Meaning of Relativity*:

> *Some try to explain Hubble's shift of the spectral lines by means other than the Doppler Effect. There is, however, no support for such a conception in the known physical facts. According to such a hypothesis it would be possible to connect two stars, S1 and S2, by a rigid rod. Monochromatic light sent from S1 to S2 and reflected back to S1 could arrive at a different frequency (measured by a clock in S1) if the number of wavelengths of light along the rod*

should change with time on the way. This would mean that the locally measured velocity of light would depend on time, which would contradict even the special theory of relativity. Further it should be noted that a light signal going to and fro between S1 and S2 would constitute a "clock" which would not be in constant relation to a clock in S1. This not only involves the loss of comprehension of all those relations which relativity has yielded, but it also fails to concur with the fact that certain atomic forms are not related by "similarity" but by "congruence" (the existence of sharp spectral lines, volumes of atoms, etc.).
(*The Meaning of Relativity*, 5th ed., p.128)

Here, Einstein offers three different arguments against SCM's redshift explanation as being caused by expanding space. First, if it were true it would mean that the speed of light must change with time, which contradicts SR. Second, light reflected between two mirrors at a constant distance can no longer be used as a light-clock because the time intervals between successive reflections would change with time. And third, the changing relationship between space and time would alter the conditions necessary for the existence of atoms, whose physical properties, including their spectral lines, depend on a fixed relationship between space and time. This contradicts observed spectral lines, which at even at huge cosmological distances remain relatively the same as locally here on Earth after adjusting for the redshift.

The SCM attempts to circumvent this objection by assuming that space does not expand within galaxies, only between them, but it is silent on how this feat is accomplished. In fact, most SCM supporters visualize the expansion as motion *in* space rather than *of* space via changing spatial metrics, thus bypassing GR and Einstein's objections. This would be consistent with the Doppler shift interpretation.

Let's now see how the SEC model will explain the redshift. It may seem that the simultaneous expansion of space and time would preserve all physics and therefore also spectral line frequencies. This would be true if two different scale-equivalent spacetimes, each of constant scales, were to be compared. However, light reaching us from distant sources in the past must work its way against a gravitational-type field created by the changing scale. Well-known from GR is that photons coming from within a gravitational field are redshifted. The cosmological scale was smaller in the past, and we may therefore informally think of ancient light as arriving from within a gravitational field induced by the scale-expansion, which would explain the redshift as a particular kind of gravitational redshift. A derivation of the redshift based on the geodesic equation for the SEC line-element may be found in Appendix I.

Hence, in the SEC, the redshift is *not* due to outward motion in space since distances to galaxies remains the same relative to co-expanding observer s during the cosmological scale-expansion except for local motions unrelated to the cosmological expansion. As we shall see, this has important implications when interpreting astronomical observations.

The Scale-expansion Idea

Let me briefly return to the scale-expansion idea since it is unfamiliar, but might be crucial to understanding the cosmos.

In fact, it might be a revolutionary new idea that will uproot and force revision of physics going all the way back to the time of Galileo. This stark implication has been an obstacle for the general acceptance of the SEC theory.

The concept of scale is of fundamental importance in our daily life; as children, the scale idea was fascinating to most of us as soon as we realized that we were smaller versions of our adult parents. Scale is a relative concept; scale is always defined in relation to something of different scale. The scale of any 3D object may be altered by changing all dimensional distances of the object by the same *scale factor*. We can make a statuette by making the scale smaller or an oversized imposing statue by making it larger.

However, we know that spatial rescaling does not mean physical equivalence. One reason is that the volume changes by the third power of the scale. Doubling the scale, making it twice as large, means that the object's volume increases by a factor of eight; the weight increases by the same factor. An object with larger scale is no longer physically equivalent to the smaller.

However, Einstein's relativity theories are based on the implicit assumption that the world is four-dimensional (4D) rather than three-dimensional, and we might ask what would happen if we were to rescale not only the three spatial dimensions but also the temporal dimension. Rescaling time means that the duration of a second will change together with the spatial scale. The atoms in any object oscillate at extremely high frequencies and thus depend on time as well as space.

In fact, if we could change the 4D scale of an object, we would find that the rescaled object would be physically equivalent to the original!

It is a theoretical fact supported by GR that the universe does not prefer any particular, or absolute, 4D scale; all scales are cosmologically equivalent.

The scale is a free parameter!

This is not too surprising, since if there were a preferred scale, something outside the universe would have to determine it, which is impossible if the four-dimensional universe is all there is.

Cosmological scale-equivalence is not only reasonable but also agrees with modern physics. Changing the metrical spacetime scale of an object by a certain constant factor will not alter Einstein's GR equations; they remain the same. This means that universes of different scales are equivalent and would appear identical relative to their inhabitants.

The cosmological scale is a hidden dimension that is the key to understanding the universe.

Cosmological Scale-expansion

In our universe, things have a scale, which is determined by the metrical coefficients of the line-element of GR (see the sidebar). A deeper discussion isn't needed here; it suffices to say that there are functions in GR that determine the scale of space and time. I will call these functions the *metrics*. Scale-equivalence in GR means that Einstein's GR equations will not change if we multiply all metrics in the line-element of GR by the same constant scale-factor. And, since the GR field equations do not change, the physical world perceived by observers everywhere in the universe, regardless of location and epoch, could be the same. Particles in diverse and distant regions would be physically the same as they are locally, which would explain the preserved relationships between spectral lines in redshifted light.

The Line-Element and Metrics

The line-element expresses the relationship between space and time by specifying the coordinate metrics as a function of location. It defines the speed of light in different directions and how the pace of time changes between spatial and temporal locations.

Although it is called a line-element, it is just a way to specify the geometry of spacetime. It is a mathematical starting point for Einstein's GR equations.

The metrics of spacetime are factors that multiply the different terms in the line-element. Generally, they are functions of both spatial and temporal location, and modify the scale for distances in the three spatial directions and the temporal direction at a certain location in spacetime. They therefore model how spacetime "stretches" or "contracts," depending on location. Gravitation is usually modeled by metrics that change with spatial location.

The scale of spacetime denotes a situation where all metrics are affected by the same single scale function. A changing scale of spacetime is therefore a simple special case of spacetime curvature. Einstein's field equations are identical for spacetimes merely differing in constant scales.

We know that the universe appears to expand—could the cosmological expansion be one of 4D scale rather than space? This would resolve a fundamental problem with the SCM by eliminating the Big Bang creation event. The universe could expand forever by changing its scale by a tiny fraction each second! Since all scales are equivalent, cosmological scale-expansion could proceed without changing the physical properties of the universe, making all epochs equivalent.

This line of reasoning leads to a new cosmological model by which the universe expands exponentially by changing its 4D spacetime scale. Here, the term *exponentially* has a mathematical meaning, which is that the scale changes by the same small fraction every second. This mode allows the cosmos to expand in proportion to itself, which also explains why existence may be eternal.

We can try to model this in GR to see what happens. GR permits us to predict the consequences of the SEC model and to compare these predictions with astronomical observations. If we can show that the SEC model better agrees with observations than the SCM, then we will have a better model.

As we will see, it turns out that the SEC model not only agrees with all observations but that it also would resolve many puzzling inconsistencies of the SCM. This should, by itself, be enough to favor the SEC over the SCM, but the power of the SEC theory does not end there; it opens the door to deeper understanding by telling us not only why the world is quantum mechanical but also why GR is incompatible with QT, which is one of the most perplexing inconsistencies of modern physics. It also tells us what causes the progression of time and suggests an explanation to the phenomenon of inertia, that is, what causes the inertial force during acceleration. And the SEC model predicts new unexpected features of the universe that may be tested; *the SEC theory is falsifiable.*

Why the SEC Model Has Been Overlooked

If these claims are valid, we might ask why the SEC model hasn't already been proposed and investigated. I can think of at least two different reasons for this.

First, an idea like cosmological scale-expansion cannot gain scientific credibility unless it is modeled by known theory, which allows us to make testable predictions. It is not enough to merely propose scale-expansion; one must also present a possible mathematical model and show that the model's predictions agree with observations. Thus, the new theory must be based on known science. This requirement is crucial because if the idea is truly new, a bridge must be built from what is already known to what's new and different. Without establishing this connection, the credibility of the new idea will suffer.

The second reason is that there is a curious side effect of education: too much of it and one may become overeducated. Albert Einstein once said with regard to this phenomenon:

"If one studies too zealously, one easily loses his pants."

A friend of mine once complained that he was "terminally overeducated." Education means that you are familiar with the teachings of modern science, including technical tools and applicable mathematics as well as accepted methods of approaching problems. Like a carpenter is limited by his tools, the modern scientist is constrained by contemporary epistemology. Actually, the scientist is in an even worse position, because the carpenter can manufacture what's needed by any means. The scientist is constrained to use accepted and approved tools. If he or she tries to use a new approach that differs from what is known, this new approach is often met with suspicion and is rejected by the scientist's peers. This is an ancient systemic epistemological dilemma in all human endeavours. Since something truly new often implies that the older thinking must be revised, new ideas are often resisted.

Scale transformation has been considered in the past. Hermann Weyl modified the GR equations in order to make them compatible with *continuous* scale transformation [Weil,1921]; and this path has been pursued and developed by several investigators—for example, Paul Dirac [Dirac, 1973]. Here, I take another approach. If I had known about this earlier work, I would probably not have developed the SEC theory. At times, ignorance is bliss.

A Puzzling Obstacle

The reason for why GR cannot accommodate *continuous* scale-expansion is rather technical, but it is easy to understand the essence of the problem if one considers scale-expansion of a time-interval like a second in relation to itself. How can we model time expanding with time? This conundrum occupied my thoughts for almost two years before I finally concluded that *it is impossible to model 4D scale-expansion that preserves all physics using standard GR.*

At first this was disappointing to me, but I did not want to give up, since I was convinced that cosmological scale-expansion ought to be physically (and mathematically) possible. I felt that cosmological scale-expansion is a beautifully simple idea, since it eliminates the need to consider a universal creation event. Also, what could possibly determine an absolute cosmological scale? Eventually it occurred to me that the fact that we cannot model scale-expansion by GR does not mean that the concept of scale-expansion is incorrect. *It might mean that GR is incomplete.*

The limitation of GR could be that it is confined to *continuous* spacetime manifolds, which means that the scale always has to progress at an even and smooth rate. However, if GR would allow discrete, stepwise scale changes, cosmological scale-expansion could be modeled quite easily. At first, this seemed to me like a capitulation, since it is very difficult to find support for a theory if it implies some shortcoming in GR. GR is the bible of modern physics and is one of its towering achievements. To suggest a problem with GR is tantamount to heresy.

But the more I struggled with this dilemma, the easier it became for me to accept that GR might be incomplete. What made me eventually accept this as a fact is that a scale-expanding cosmos not only agrees with observations and resolves a number of problems with the SCM, but it would also settle the two most vexing problems of modern physics: it would explain the nature of the progression of time and provide the missing connection between GR and QT.

During my years of working on the SEC theory, I held on to the following *guideline*:

> *The cosmos is scale-equivalent. Therefore, the cosmological expansion may be 4D scale-expansion. We should be able to model this expansion mode mathematically.*

By the SEC model, the duration of a second increases by the minute fraction $1/T$ each second, where T is the Hubble time (see sidebar), which is about 14 billion years = $4.2 \cdot 10^{17}$ seconds. This also is taken to be the age of the SCM universe.

Hubble Time and Cosmological Distance

Hubble Time is the apparent time to the Big Bang assuming the redshift is a Doppler effect. Currently it is estimated to be about 13.8 billion years, and is believed to be the age of the universe. The Hubble distance is obtained by multiplying the Hubble time by the speed of light. In the SEC model, Hubble Time is a constant unrelated to the age of the universe. The value of Hubble Time depends on the value of the Hubble constant, which relates redshift to distance.

Measurement of a galaxy's redshift is relatively easy, but accurately determining its distance is not that easy. The traditional method of determining distance is to find a "standard candle," which is a star or galaxy of a known brightness. If such a star can be found that is of a known intrinsic brightness, its distance can be figured out by determining its apparent, measured brightness, since the brightness falls off (roughly) as the inverse square of the distance.

One such standard candle is the Cepheid variable type of star. These stars brighten and dim with regular periods. Their intrinsic brightness has been found to be the same if their period is the same.

NASA sponsored a long-term program to find and measure Cepheid variables in 18 galaxies using the Hubble Space Telescope. In May of 1999, after eight years of data gathering, it was announced that the Hubble constant was 70 kilometers per second per kiloparsec (one parsec=3.26 light years) within an uncertainty of 10 percent. This translates into a Hubble Time of 12 to 13.5 billion years. Another group led by Allan Sandage immediately challenged this announcement. They determined the Hubble Time to be 14 to 18 billion years based on more than 30 years of ground-based observations.

They used not only Cepheid variables, but also other standard candles such as a class of supernovae, whose intrinsic brightness is determined from the time it takes for them to brighten and fade out.

In June of 1999, a group at the National Radio Astronomy Observatory announced a new method of measuring galactic distances. Instead of using standard candles, this method uses very long baseline interferometry radio astronomy from many widely separated antennas. The result is a spatial resolution about 500 times better than the Hubble Space Telescope, but at invisible radio wavelengths. It measured the speed of orbital motion of a natural maser (or radio hot spot) orbiting a galaxy (NGC 4258) and determined its location at two different times relative to the galactic center.

Then the distance to the triangle formed by these three locations was determined using straightforward trigonometry with an uncertainty of about 4 percent. This was one of the 18 galaxies whose distance previously had been determined by the Hubble Space Telescope using the Cepheid variable method. The Cepheid variable method calculated the distance to NGC 4258 as 28 million light-years, whereas the radio astronomy calculation was 23.5 million light-years. The new method when applied to this single galaxy produces a Hubble Time of about 10 billion years. In the future, hot spots found orbiting other galaxies should yield even better estimates of the Hubble Time.

By GR the SEC is modeled by a cosmological line-element with scale-factor $exp(2t/T)$, where t is Atomic Time and T the Hubble time. It predicts that the frequency of light decreases exponentially with time and distance:

$$f = f_0 \cdot e^{-t/T}$$

This frequency–distance relation models the same fractional decrease in frequency (and photon energy) for every second of light travel time. Thus, a photon loses energy with time and distance and becomes red and "tired." (However, the SEC redshift is not the same as the traditional tired-light redshift model because there is additional dimming of the light in the SEC, just like in the SCM. I will return to this later.)

How Tired Light Explains the Redshift

The Doppler Effect is not the only way in which light can be redshifted. Expanding space and time will cause a redshift that depends on the distance light travels rather than on the velocity of a receding source.

When photons move through expanding spacetime, they gradually lose energy. The frequency becomes lower and the light is shifted toward red in the spectrum. This type of redshift has been considered in the past and is usually referred to as *Tired Light*.

Tired Light seems to agree with observations better than the Doppler Effect. Some have tried to explain it by assuming that the photons collide with other particles on their way and therefore lose energy. However, collisions would cause the photons to change direction, which ought to make the image of distant galaxies fuzzy like a streetlight in fog.

In the SEC theory, the Tired Light redshift–distance relation is caused by uniform expansion of spacetime. It is a gravitational-type effect caused by curved spacetime, which in the SEC is induced by the cosmological expansion.

This redshift relation also provides a simple relationship between redshift and distance, because we know that light moves with velocity, c:

$$\text{Distance} = d = c \cdot t = c \cdot T \cdot \ln(\frac{f_0}{f}) = c \cdot T \cdot \ln(z+1)$$

Here, the z indicates the magnitude of the redshift. Locally, in the absence of redshift, we have $z = 0$. The redshift z is positive for distant light and increases with increasing distance.

This relationship between distance and redshift greatly simplifies the interpretation of astronomical measurements in the SEC, which in the SCM is a quite complicated affair since the interpretation depends on several unknown parameters, which often are juggled to obtain the best fit to observations. In the SEC, there are no other parameters than the Hubble time, T, yet as we shall see in the next chapter, the agreement with observations is superior to that of the SCM.

Obviously, the cosmological redshift is interesting, but it pales in comparison to a new, and totally unexpected, feature of the SEC—*cosmic (velocity) drag.*

Cosmic Drag

Based on GR we can show that relative velocities of freely moving objects and particles diminish with time in the SEC (in what follows I will take a 'particle' to mean an object with positive rest mass regardless of size). For velocities much less than the speed of light we have a relation similar to that of the redshift:

$$v = v_0 \cdot e^{-t/T}$$

The velocity of a freely moving particle decreases exponentially with time. This also applies for angular momentum, which diminishes in a similar way. This is what the *geodesic equation* (see sidebar) of GR predicts, and it should be possible to confirm this new and unexpected phenomenon from observations taken in the solar system. If there is cosmic drag, Newton's first law of motion no longer holds, which would be an important discovery with revolutionary consequences. Galileo and Newton would be wrong, and Aristotle, who thought that all motion diminishes over time, would be right. But, as already mentioned, the relativistic momentum is preserved for freely moving particles if we take into account the expanding temporal metric.

If particles slow down, wouldn't this time-diminishing velocity also apply to photons? No, a particle that initially moves at the speed of light in the SEC will not slow down but keep moving at the speed of light, see Appendix I. However, because the energy of a photon is proportional to its frequency a photon loses energy as given by the redshift relation.

Geodesics

Geodesics are trajectories followed by very light freely moving particles computed using the geodesic equation of GR. In the absence of spacetime curvature, these trajectories are straight lines, while in gravitational fields, they may be curved. A particle that does not fall in a gravitational field is subjected to a gravitational force. This is what causes the weight of objects on the surface of the Earth. In a gravitational field, a falling object is in a situation similar to an object at rest in outer space; there is no gravitational force. See further Chapter 5.

A Cosmological Reference Frame

The existence of a cosmological reference frame has been hotly debated ever since the time of Newton and Leibniz. Remember that in Newton's celebrated bucket experiment, he observed that the surface of the water in a spinning bucket becomes concave and concluded that the water in the bucket somehow "senses" that it is spinning. This is, of course, caused by the centrifugal force, but what generates this force? It is interesting that ever since Newton this simple question has gone unanswered. It is not a local effect here on Earth; the same force acts on the planets in the solar system and on stars in galaxies.

Would the same force exist for a spinning object in a totally empty universe? Einstein asked himself if an orb of water held together by gravity would assume an oblate form when spinning in an empty universe due to centrifugal acceleration. He wondered how we would know if there is rotation in a totally empty universe.

Ernst Mach, a famous German physicist/philosopher in the late 1800s (the Mach number, which is the ratio between velocity and speed of sound, is named after him) did not think we could; he thought that matter far away in the universe somehow affects local motion. This is Mach's principle (see sidebar). Einstein was influenced by these ideas, which inspired him in his struggle with GR. But, unfortunately, GR does not explain the centrifugal force or the closely related inertial force.

Mach's Principle

Mach's Principle is an alternative to Newton's absolute space. According to Mach's Principle, the fixed reference space is defined by the presence of distant matter in the universe. Einstein was greatly influenced by this idea when he developed his theory of general relativity, hoping that this theory also would solve the mystery of inertia. However, he found that this was not the case.

An "inertial frame" is a coordinate frame that moves at a constant velocity and direction in the absence of gravitation. Classical physics, as well as relativity, is insensitive to this motion; all inertial frames are physically equivalent. We don't feel velocity, but we feel acceleration.

Although inertial and centrifugal forces exist, there does not seem to be any absolute reference frame; that is, nothing appears to be fixed relative to which acceleration or rotation may be measured. Perhaps something important is missing in our understanding, and this has been a central conundrum since the beginning of modern physics.

I have already mentioned that the relative galaxy velocities are unexplainably small. Numerical simulations of the motion of galaxies during cosmological expansion under the influence of mutual gravitational attraction show much higher velocities than are observed. And, the Cosmic Microwave Background (CMB) radiation is not uniform in all directions; there is a so-called dipole, which indicates that we are moving at about 370 km/s relative to a uniform CMB background. How can this be possible if there is no reference frame?

This fundamental conundrum is resolved by the SEC theory:

The cosmological scale-expansion induces a cosmological reference frame.

Because the cosmological expansion acts everywhere, this cosmological reference frame exists everywhere. Cosmic drag slows *relative* velocities and dissipates *relative* rotations. Thus, cosmic drag defines a spatial reference frame as being the frame toward which all motion converges. *This reference frame is being actively maintained by diminishing relative motion.* In principle, it is possible to find this absolute frame as being the frame relative to which all relative motions add to zero. As a consequence, the centrifugal force in Einstein's thought experiment would be present even in an empty scale expanding universe without matter.

Newton presumed the existence of an absolute space without explaining it; it simply was part of God's creation. Mach relied on distant matter in the cosmos as defining the reference frame for all motion, but did not explain how very distant matter could influence local matter or why distant stars (galaxies) appeared to be stationary. With standard physics, finding any reasonable explanation to why there should be a cosmological reference frame is impossible. Yet, everything we know tells us that a cosmological reference of rest exists, and this was also the consensus at the beginning of the twentieth century.

When Einstein's SR theory appeared, it muddled the water. Don't get me wrong; I am an ardent Einstein admirer, since without him there would be no SEC theory. But he could not have guessed that relative velocities might diminish over time (unless they initially equal the speed of light). This follows from the geodesic equation for the SEC line-element, see Appendix I.

Einstein made the very natural assumption that since the velocity of light by Newton's first law is independent of a reference frame, this should also apply to all other velocities and therefore there should be no preferred cosmological reference frame.

In the SEC, the reference frame is created by cosmological expansion in a fifth dimension—the cosmological scale. Although perhaps difficult to visualize, expansion of all four metrics might occur even in the absence of matter, and thus, the cosmological reference frame could exist even in a vacuum without matter. The expanding scale acts everywhere across the cosmos at large distances. It slows all relative motion.

The cosmological scale-expansion explains the cosmological reference frame.

Also, cosmological scale-expansion would induce spacetime energy of a different kind than what is found in matter or radiation. This is the missing "Dark Energy". Therefore, the cosmos would not be empty even in the absence of matter and radiation.

On a more philosophical note, we might perhaps wonder why there is cosmological scale-expansion in the first place. Although we now know that it appears to be compatible with the cosmos we observe, is there more to it?

According to Einstein's field equations, the metrics of spacetime are defined by a given energy-momentum tensor (except for a constant scale factor), which seems to imply that the metrics should remain undetermined if the energy-momentum tensor disappears. This was also Einstein's initial objection when he learned about Karl Schwarzschild's [Schwarzschild, 1916] exterior solution of his equations, which models a black hole, since it implies that the spacetime geometry is well defined even infinitely far away from matter. He wondered what possibly could define the metrics of spacetime in the absence of energy.

From this point of view, the existence of metrics, and therefore of spacetime itself, seems to demand the existence of an energy-momentum tensor. In other words, spacetime would not exist without cosmological scale-expansion, which induces vacuum energy.

We might say that expansion of the cosmological scale sets the stage for all existence; without it, nothing would exist.

This is also consistent with the implication that the ever-slowing progression of time is the ultimate and perpetual energy source for the universe.

We exist because the cosmos expands.

If the cosmos did not expand we would not be here to wonder why it expands!

CHAPTER 4

Problems with the SCM and Their Resolutions

Science alone of all the subjects contains within itself the lesson of the danger of belief in the infallibilityofthegreatestteachersinthepreceding generation.... Learn from science that you must doubt the experts. As a matter of fact, I can also definescienceanotherway:Scienceisthebeliefinthe ignorance of experts.

—Richard Feynman

Observational Verification of the SEC Theory

I will now justify the SEC theory by presenting arguments and observational findings that cannot be ignored. By the rules of scientific exploration, a theory that better agrees with observations should be preferred. This rule is also known as Occam's razor. This razor is sharp and unyielding;

it cuts out, and discards, competing theories that cannot pass muster regardless of how well accepted they are.

Cosmological Tests

Several observational programs have been designed with the main objective of testing cosmological models. One might think that if the SCM's predictions disagree with these tests it ought to defuse enthusiasm for the theory. However, this has not been the case. Support is as strong as ever for the SCM despite several undisputable observational discrepancies. This is quite ironic, if not disingenuous, since the cosmological tests originally were designed with the primary objective of disqualifying models that do not agree with observations. However, instead of rejecting the SCM, its supporters keep adding new speculative (mythological) features to this model in order to explain away the discrepancies. And in spite of these discrepancies, popular expositions of the SCM theory are encouraged and presented to a gullible public in books and TV programs.

Several investigators beginning with Edwin Hubble (see sidebar) have argued that astronomical observations agree better with the tired-light redshift model than with the Doppler-like redshift of the SCM.

Edwin Powell Hubble

Hubble was born in the small town of Marshfield, Missouri, USA, on November 29th, 1889. In 1898, his family moved to Chicago, where he attended high school. Young Edwin Hubble had been fascinated by science and mysterious new worlds from an early age, having spent his childhood reading the works of Jules Verne (*20,000 Leagues Under the Sea, From the Earth to the Moon*) and Henry Rider Haggard (*King Solomon's Mines*). Edwin Hubble went to Oxford University on a Rhodes scholarship, where he did not continue his studies in astronomy but instead studied law. At this point in his life, he had not yet made up his mind about pursuing a scientific career.

In 1913, Hubble returned from England and was admitted to the bar, setting up a small practice in Louisville Kentucky; but it didn't take long for Hubble to realize he wasn't happy as a lawyer, and that his real passion was astronomy. So he studied at the Yerkes Observatory, and in 1917, received a doctorate in astronomy from the University of Chicago.

Following a tour of duty in the First World War, Hubble took a job at the Mount Wilson Observatory in California, where he took many photographs of Cepheid variables through the 100-inch reflecting Hooker telescope, proving they were outside our galaxy and determining the existence of several other galaxies such as our own Milky Way.

Hubble had also devised a classification system for the various galaxies he observed, sorting them by content, distance, shape, and brightness. It was then he noticed redshifts in the emission of light from galaxies, which seemed to increase in proportion to their distances. From these observations, he was able to formulate Hubble's Law in 1929, helping astronomers determine the age of the universe and suggesting that the universe was expanding. Hubble's law states that distances to galaxies are proportional to their redshifts.

Source: www.edwinhubble.com/hubble_bio_001.htm.

An important paper by Paul LaViolette [LaViolette, 1986] presents clear observational evidence showing that the tired-light redshift–distance relation (which is the same as in the SEC) agrees with cosmological tests without resorting to any of the many speculative evolutionary scenarios needed to reconcile the observations with the SCM. But, unfortunately, this significant contribution has largely been ignored. Since 1986, our observational capabilities have improved dramatically with new tools like the Hubble space telescope and Very Long Baseline Interferometry (VLBI), and as we shall see, it has gradually *become clear that the SCM simply does not agree with the observations.*

In the following, I discuss three standard cosmological tests: the galaxy number count test, the angular size test, and the surface brightness test. Also, recent supernovae observations are examined.

The Galaxy Number Count Test

Any candidate cosmological model should be able to predict how the number of galaxies (galaxy count) we see from the Earth increases with distance. Since the apparent luminosity of galaxies (how bright they appear) depends on their distance, there is a corresponding test for number count as a function of apparent luminosity. However, several observational programs have repeatedly found that the *SCM's predictions do not agree with observational data.*

Figure 5 shows a summary from sixteen different number count programs taken from a paper by Metcalf et al. (1995). Galaxies seen in the sky within a spatial square degree and a 0.5 magnitudes luminosity range are counted and displayed as a function of luminosity magnitude. The

SEC theory's prediction has been added to a figure presented in the Metcalf paper. It is clear that the SCM model clearly fails the test, since its graph lies well below the observations, while the SEC model agrees well with the observations.

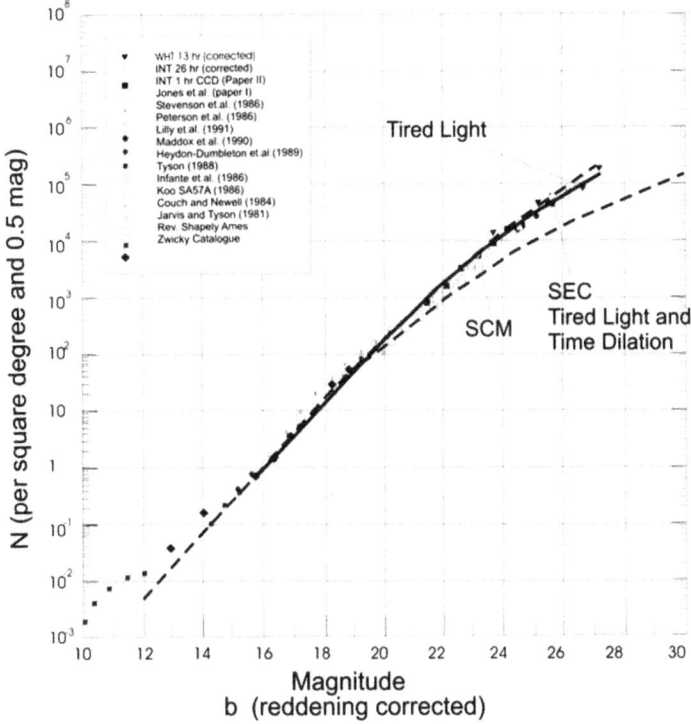

Figure 5: Galaxy number counts

The SEC data points in the plot were obtained using the distance-redshift relation:

$$Distance = c \cdot T \cdot \ln(z+1)$$

Here, z is the redshift; c, the speed of light; and T, Hubble time. This relation has one free parameter: Hubble time.

Metcalf et al. [Metcal et al, 1995] try to explain this clear discrepancy between the SCM's prediction and the observations by proposing several evolutionary scenarios, but none of these fits the data as well as the SEC. This is firm observational evidence in favor of the SEC model.

In figure 5 the magnitude displayed on the x-axis indicates apparent luminosity of galaxies; the larger the magnitude is the dimmer is the galaxy. The magnitude scale is logarithmic and a difference of one in magnitude corresponds to a factor 2.512 dimmer luminosity. As seen by the human eye, visible stars have luminosities less than 6. Therefore, galaxies at magnitude 26 are 100 million times dimmer than what we can see with our naked eye, which explains why we have to use the Hubble space telescope and charge-coupled devices to see them. The y-axis indicates the number of galaxies observed within one square degree of the sky and a 0.5 magnitude band.

The Angular Size Test

The angular size of a cosmological object—for example, a galaxy, may also be used to test candidate models. The SCM predicts that beyond a certain distance, the angular size will start to *increase* with increasing distance rather than decrease; in the SCM there is a minimum in the graph of angular size versus distance. However, observations do not support this; they show that the observed angular sizes decrease monotonically with increasing distance.

Figure 6 is from a paper by Djorgovski and Spinrad [Djorgovski and Spinrad, 1981]. The SEC prediction has been added. Here, the SEC distance–redshift relation is used again. The observed angular size should, on the average, be proportional to the inverse of this distance, which is the graph for the SEC model in figure 6. Clearly, the SEC model's agreement with the observations is superior. There is no evidence that the angular size has a minimum value.

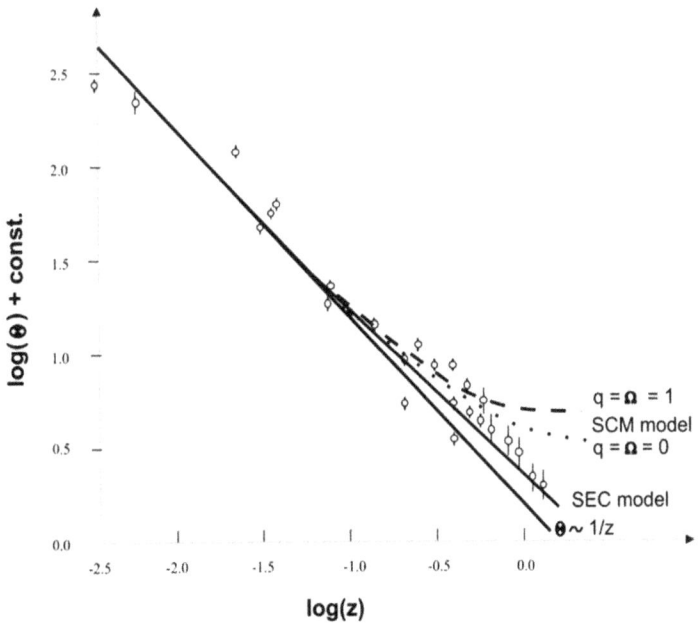

Figure 6: Angular size observations

For small redshifts the distance is proportional to z, which is the lower line in the figure. However for the SEC model, the distance is proportional to ln(1+z), which is shown as the middle line. The distance relation for the SCM is more complicated, *because it depends on parameters like the mass density parameter, Ω, and the deceleration constant, q. Two cases for the SCM are shown. These curves indicate that the angular size should decrease more slowly with distance in the SCM model, which partly is due to the fact that the maximum distance in the SCM universe is the Hubble distance, which in the SCM occurs at infinite redshift. However, in the SEC model, the Hubble distance is reached at $z = 1.7$. Clearly, the data supports the SEC.*

The Surface Brightness Test

The surface brightness test is a powerful and robust discriminator between candidate cosmos theories. Brightness is defined as observed luminosity per surface area of the observed object. Usually, surface brightness of galaxies is estimated based on apparent luminosity per observed surface area as measured in squared arc-seconds. The SCM fails this test since its predicted values do not agree with observations.

Observational results reported by Lubin and Sandage {Lubin and Sandage, 2001} show that the SEC theory agrees with observed galaxy surface brightness while the SCM does not. The solid line in figure 7 is the calibrated surface brightness baseline estimated from nearby galaxies. The SEC theory's predictions agree well with the local surface brightness (filled symbols). However, there is disagreement with the SCM as shown by the heavier, outlined open symbols.

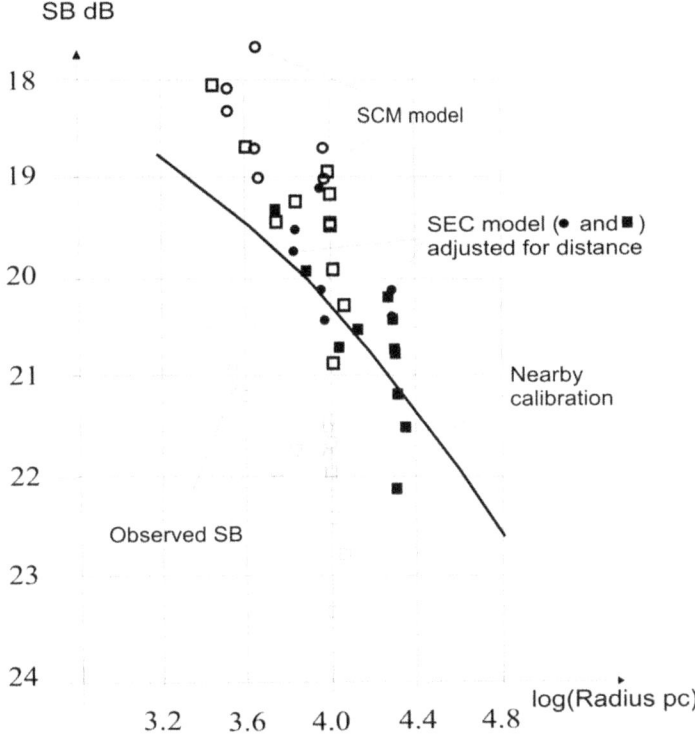

Figure 7: Surface brightness observations

The lower left part of the graph shows the observed surface brightness of galaxies in two clusters at redshifts close to one (z = 1) as represented by the faint open squares and circles. The x-axis shows the logarithms of the estimated radii of these galaxies, and the y-axis the surface brightness magnitudes. Observations of nearby galaxies are represented by the solid curved line. Any candidate cosmological model can be used to predict what the surface brightness of a distant galaxy would have been if it instead had been located nearby. The solid open symbols represent the SCM model's predictions and the filled symbols the SEC model's predictions. The *figure shows that the SEC model predicts surface brightness of distant galaxies that, on the average, are the same as for nearby galaxies. On the other hand, according to the SCM model, distant galaxies were brighter, a discrepancy that typically is explained away by evolution.*

Again, the agreement between observations and the SEC model's predictions must be acknowledged; distant galaxies appear to have the same surface brightness as nearby galaxies.

Accelerating Cosmological Expansion?

A startling recent finding that contradicts the SCM is based on supernovae observations, in particular a certain type of supernovae—the so-called supernovae (SN 1a with plural SNe 1a).

In this scenario, a carbon-oxygen rich white dwarf star is accreting matter from a companion star. (The kind of companion star that is best suited to produce type 1a supernovae is hotly debated.) In a popular setting, so much mass accumulates on the white dwarf that its core reaches a mass density of 2×10^9 g/cm^3. This causes uncontrolled fusion of carbon and oxygen, thus detonating the star.

This is believed to be a repeatable process, always resulting in a characteristic radiation signature. I will not go further into these details more than to say that the light-curve from an SN 1a is closely related to its light output. The light-curve is a graph of its luminosity as a function of time that typically could last a month or longer, and its intrinsic luminosity may be estimated from the shape and duration of this light-curve. Also, the spectrum of an SN

1a can be recognized and distinguished from other types of supernovae.

The longer the duration of the light-curve, the brighter is the supernova, which makes it possible to use them as "standard candles." Since we can measure the apparent luminosity and can use the light-curve to estimate the emitted, intrinsic luminosity, we can use this information to estimate its distance and test the validity of different cosmos models. Furthermore, since the light output is enormous, often greater than that of a typical galaxy, and since they flare up and die away over a couple of months, the SN 1a can be detected against the background light (typically a host galaxy) and give information on geometrical properties of the very distant universe.

SN 1a observations have given important but unexpected information. It appears that the cosmological expansion is now *accelerating*. This was stunning news and the *Science* magazine called it the breakthrough of the year in 1998 [Glanz, 1998]. An accelerating universe suggests a new force, possibly implying a cosmological constant as originally proposed by Einstein, and this force could be related to the missing dark energy predicted by the Inflation theory. This was exciting news and deserved the breakthrough nomination.

However, this interpretation crucially depends on the redshift–distance relation of the SCM. As we shall see there is another, better, interpretation to the SNe 1a observations.

The recently reported SNe 1a observations by the Supernova Cosmology Project [Perlmutter 2003; Perlmutter et al. 1995, 1997, 1999] and by the High-Z

Supernova Search Team [Schmidt et al., 1998] seemingly confirm that these observations do not agree with the SCM unless the cosmological expansion accelerates. However, as shown in figure 8 below, the SNe 1a observations agree very well with the theoretical predictions of the SEC model. This good agreement with the SEC model is obtained without adjusting any parameters.

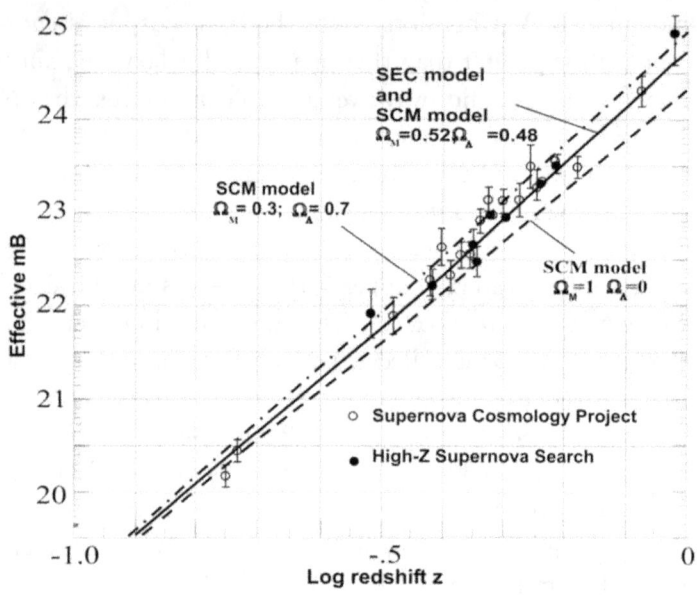

Figure 8: SNe 1a observations

The lower dashed line is the prediction of the SCM if the cosmological mass-density equals Einstein's Critical Density. The upper dashed curve is the SCM with 30 percent mass density and 70 percent cosmological constant, which together makes $\Omega = 1$. This interpretation is currently favored by the SCM supporters. The solid line is the SEC model's prediction.

Note that the SCM model's predictions assume that the energy density equals Einstein's Critical Density for which $\Omega = 1$, which is about twenty times larger than the visible mass-energy density in the universe. The missing 95 percent is believed to be a combination of dark matter (about 30 percent) and dark energy (about 65 percent) of unknown origin. This is one of the most puzzling unresolved problems in contemporary cosmology. On the other hand, since the SEC model's predictions agree with observations, no energy is missing. In fact, what in the SCM appears as missing, unexplainable energy is in the SEC energy contained in spacetime itself; it is induced by the cosmological scale-expansion! This directly follows from the vacuum energy-momentum tensor for the SEC theory, see further below.

An even more recent paper by Riess et al. [Riess et al, 2004] presents data for 16 newly discovered SNe 1a, 6 of them observed in the redshift range $z > 1.25$ using the Hubble space telescope. These new observations suggest that the universe initially went through a phase with a *decelerating* expansion rate, which later was followed by *accelerating* expansion. Riess et al. modeled the evolution of the luminosity distance by assuming an initial phase with a linearly decreasing positive deceleration constant, later followed by an accelerating phase. The parameters of this model as well as the redshift at which the deceleration constant is zero may be estimated from observational data and be used to model the evolution of the luminosity distance. The transition from decelerating to accelerating expansion is estimated by Riess et al. to occur at $z = 0.46$. Our figure 9 (which is figure 4 in Riess et al. with the SEC prediction added) shows the fit to the observations assuming flat SCM cosmology with $\Omega_M = 0.29$ and $\Omega_\Lambda = 0.71$.

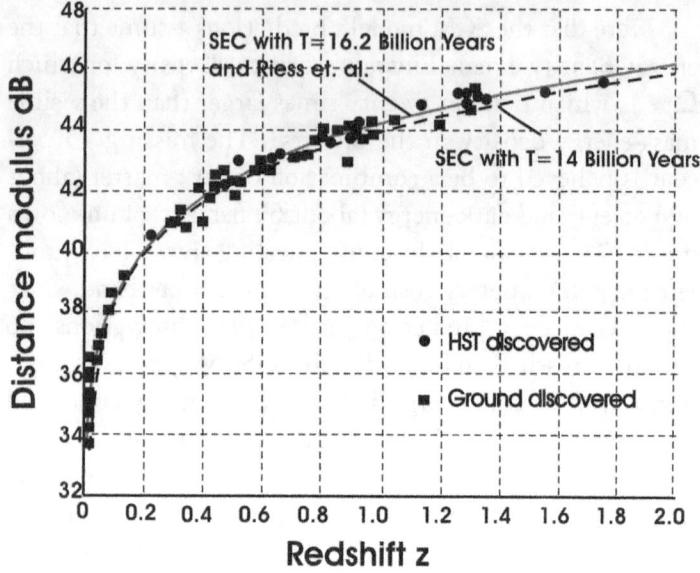

Sne Ia match to the SEC luminosity distance relation compared to the fit by Reiss et.al.

Figure 9: SNe 1a distances vs. redshifts

Here the redshift is plotted on the x-axis and the distance modulus in magnitude, as predicted by the cosmos model used, is on the y-axis. The distance modulus is proportional to the logarithm of the predicted distance to a source expressed as a function of its redshift. Again, the agreement with the SEC prediction is obtained without any adjustments.

The SEC theory's prediction may be compared to the curve in figure 4 in the paper by Riess et al. If the Hubble distance is D=16.2 billion light-years there is no difference whatsoever between the two curves over the whole range from z=0 to z=1.8. The SEC graph for D=14 billion light-years is very close with a slightly better fit at higher redshifts.

This excellent agreement provides strong support for the SEC theory without requiring additional speculation on cosmological acceleration or deceleration.

Note: HST = Hubble Space Telescope.

Summarizing the observational evidence:

Three cosmological tests and the recent SNe 1a observations all agree with the SEC model, while the SCM model's predictions disagree with all these observational tests. Clearly the SEC ought to be favored over the SCM.

Explaining Galaxy Formation and the Flat Rotation Curves

Further support for the SEC model may be obtained from the shape and velocity profiles of spiral galaxies. It turns out that it is very difficult to explain spiral galaxies in the SCM scenario because they defy known physics. Computer simulations have been performed in which large numbers of particles are allowed to move under the influence of mutual gravitational attraction, simulating stars in a galaxy. However, these simulations do not look at all similar to the galaxies we see. The main obstacle is what commonly is referred to as the *angular momentum problem*. A body, such as the Earth, which moves around another body (the Sun) will, according to standard physics, continue to move at the same speed in the same orbit forever, assuming no frictional or other influences. This is because the angular momentum is conserved. The same applies to stars in a galaxy; since they cannot shed angular momentum (although it might be transferred between them by collisions and gravitational actions), the stars will continue to move without settling down in the nice spiral shapes we see in our telescopes.

Galaxy Formation

Given a certain initial distribution of tiny particles (dust) in space, we can compute how long it will take for the dust to condense into stars under the pull of gravitation, and the additional time that would be required for these stars to congregate into galaxies. Furthermore, in order to create the universe we actually see, galaxies have to form galaxy clusters, and these clusters form sheets and filaments on a scale of hundreds of millions of light-years. The time required for all this to happen is hundreds of billion years, an order of magnitude longer than the Big Bang age of the universe of 10-14 billion years. This is an acute, unsolved problem for the Big Bang theory.

There is also another problem referred to as the *flat rotation curves*. If one looks at a spiral galaxy on edge, one sees a narrow, glowing band of stars with a bright bulge in the middle. Stars on one side of the bulge move toward us and on the other side away from us, since they move in circular orbits around the bulge. When we measure their velocities from Doppler shifts in their spectral lines, we find that the velocities are fairly constant independent of the radial distance from the galaxy core. This is surprising since by standard physics we would expect them to decrease farther out from the center if most matter is confined to the middle as indicated by visible stars. It appears that there is *dark matter* that we cannot see, perhaps a large amount of matter in the form of invisible gas. To explain this, the assumption is made that each galaxy is surrounded by a "halo" of dark matter (which is different from the cosmological dark energy). But even postulating the existence of dark, invisible halos is not enough to explain the formation of galaxies; these dark halos are assumed to somehow also absorb the angular momentum of visible matter. This is, of course, quite speculative and clearly indicates that we simply don't understand what is going on. The galaxies we see cannot be explained by standard physics unless we add dark matter and some explanation for the shedding angular momentum. This is yet another example of an attempt to paper over a problem with the SCM.

In the SEC, cosmic drag explains the loss of angular momentum. Spiral galaxies might form when matter freely falls on spiral paths toward the galaxy core, which would permit the formation of galaxy arms via gravitational attraction. This explanation certainly is superior to assuming that there is dark matter of unspecified nature that somehow absorbs angular momentum, which clearly is a solution that is forced by the lack of alternatives.

Galaxies could be very old in SEC, perhaps tens or even hundreds of billions of years old. They are dynamic objects, with matter continuously falling toward the galaxy core due to cosmic drag. From time to time, matter (gas) might be ejected from the galaxy core in order to keep the matter inside the central bulge constant on the average. This could be the role of the active galactic nuclei (AGNs) and "black holes," which have been discovered at the center of most galaxies. In the SEC context black holes cannot form. Therefore some mechanism must exist that prevents the formation of a black hole; see further the sidebar.

Black Holes

The SCM general relativity relations (assuming a vacuum energy-momentum tensor with zero components) predict that nothing can prevent a gravitational collapse if the mass density becomes high enough. It is generally believed that a star several times larger than the Sun might, after all nuclear fuel has been expended, collapse under gravitation, and reach a state where nothing can prevent further collapse into a singularity of infinitely small size. This is a black hole, which gets its name from the fact that gravitation is so strong in its vicinity that not even light can escape its pull. Obviously, if black holes really exist they are very strange objects indeed. Although people speculate that galaxy cores may contain black holes, nobody has been able to prove that they really exist.

In the SEC theory, the energy-momentum tensor in vacuum is not zero, and as a consequence, black holes are not possible according to Einstein's field equations.

Stars moving on geodesics will follow spiral paths toward the galaxy core, and since they are freely falling, gravitation pulls them into the shape of galaxy arms, resulting in the well-defined spiral arm structure and the thin galaxy discs we see.

Since stars are freely falling toward the center of a galaxy during billions of years, gravitation has time to pull matter together into galaxy arms with increasing mass density closer to the core. In the outer region of a galaxy, the matter is mostly hydrogen gas; but closer in, where the mass density becomes higher, star formation begins and the arms become visible. This explains why younger stars are usually found farther away from galaxy cores, while the central region of galaxies mainly contain older stars, judging from their spectral signatures. The Galactic Explorer Telescope (GALEX) satellite launched in 2003 has provided stunning confirmation that new stars are formed in the outer regions of galaxies (available at the GALEX Web site, www.seds. org/messier/more/m101_galex.html). The second picture at this web site clearly shows bluer new stars far away from the core and that stars are getting progressively redder and older closer to the core. This is consistent with the prediction that stars in a spiral galaxy are approaching the core over time due to cosmic drag. The front cover of this book also shows that the stars are older closer to the core.

The SEC model may be used to predict the motions of a typical star in a galaxy. Figure 10 shows predicted galaxy shapes compared to two actually observed galaxy shapes. Note that a spiral is not the trajectory followed by a star but is formed during numerous rotations around the galaxy center. A spiral galaxy rotates like a giant wheel.

Appendix III analyzes the motion of stars in a galaxy in view of the SEC model and derives both the spiral orbits of stars and their flat velocity curves. I show that both these properties are natural consequences of the SEC model's spacetime, which is curved by the scale-expansion. *Newton's laws of motion no longer apply.*

With the SEC model there is no need to assume that each galaxy is surrounded by an invisible halo that somehow absorbs angular momentum because the observations agree with the SEC theory. The amount of invisible Dark Matter could be much less than the currently estimated five to ten times the visible matter.

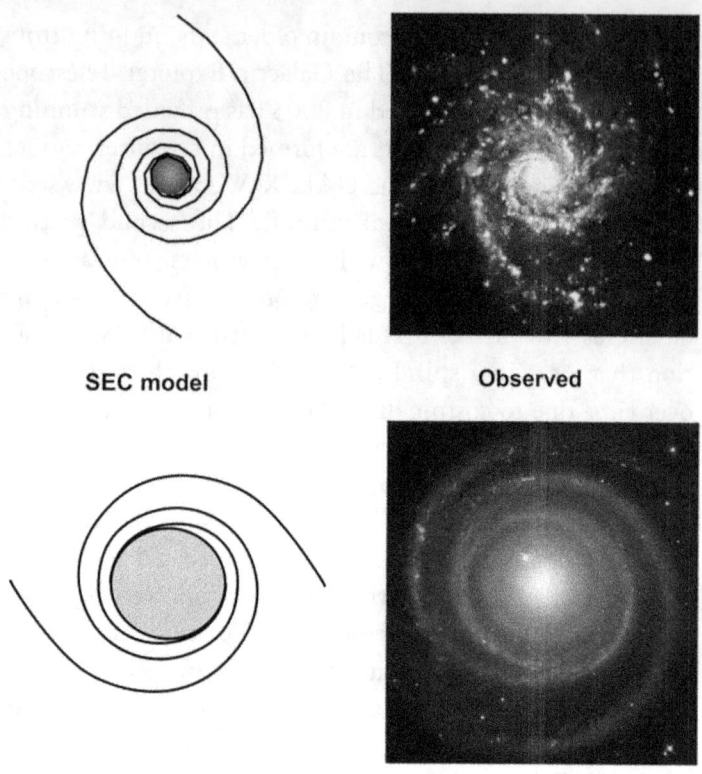

SEC model   Observed

Figure 10: Modeled and observed spiral galaxies

Philosophical and Conceptual Problems with the SCM

Many believe that the SCM gives a very good description of our universe. Occasionally, one problem or another is mentioned, but generally, the impression is given that all is well with the SCM model and that only minor adjustments remain. However, as we have seen, this is far from true; *the SCM model's predictions disagree with observations.* But this is not the only problem because there are also conceptual difficulties with the SCM. In the following, I show how the SEC model often quite easily resolves these puzzling dilemmas.

Creation of the Universe

A fundamental philosophical and physical difficulty with the SCM is, of course, the beginning of the world in which the universe was supposedly created instantaneously. This idea is unpalatable because it implies the breakdown of physics at the time of creation, which would make the origin of the universe forever incomprehensible. Identifying an alternative explanation that could address the origin of the universe while staying within the bounds of physical laws would therefore be quite attractive.

Tracing the origin of the cosmological creation idea, which is deeply rooted in Western society, is interesting. The creation myth may be traced at least as far back as to the cradle of Western civilization in Mesopotamia some 6,000 BC, and the theme of creation also runs strong through ancient cosmogonies elsewhere. Christian scholars inherited the creation idea from Jewish teachings. People in the West have long believed that God created the world, and this belief is still strongly held by many in our modern society.

Perhaps we might feel a bit superior when reading ancient creation stories because of their often fanciful mythological descriptions. But, if this is the sole reason for rejecting them, we have already committed a fundamental mental error by presuming that the universe was ever created! Although it seems so obvious to us that the universe must have had a beginning at some time in the very distant past, this idea might be nothing more than a deep-seated misconception.

Parmenides of Elea, who lived around 500 BC, clearly recognized this. Here is his celebrated insight:

> ONLY BEING IS — NON-BEING IS NOT. BUT, IF ONLY BEING IS, THERE CAN BE NOTHING OUTSIDE THIS BEING THAT ARTICULATES IT OR COULD BRING ABOUT CHANGE. HENCE, BEING MUST BE CONCEIVED AS ETERNAL, UNIFORM AND UNLIMITED IN SPACE AND TIME.
>
> —PARMENIDES

I think that this simple but profound observation is logically irrefutable; yet fully embracing it might be difficult for most of us, since it challenges our worldview and, as we shall see, also the foundation of modern science. Parmenides argued that either the universe exists or it does not exist. If it exists it cannot have been created from nonexistence since, by definition, there can be no nonexistence.

Therefore, he concluded that the cosmos must always have existed.

Unfortunately, the belief in the creation, which was supported by religious dogma, prevented Parmenides' ideas from gaining foothold some 2,500 years ago.

The SCM might well be a modern version of the creation myth. Not long ago, any challenge to the creation idea amounted to heresy, and religious, as well as emotional, resistance to the idea of eternal existence ran strong— and still does.

However, this resistance is not based on science.

Some say that the world was created in a truly gigantic vacuum fluctuation. But this assumes that space and time existed before this fluctuation took place, since quantum mechanics cannot exist without the existence of time and space (and as we shall see, the scale-expansion). We might ask for how long spacetime existed before this vacuum fluctuation occurred, or was perhaps everything created from nothingness at exactly the same instant in time (still spacetime must have come first!)? Obviously, science cannot, and will never be able to answer this question. Therefore, the Big Bang creation scenario violates known physics and is in this respect as unscientific as any ancient creation myth. It is therefore strange that the SCM has gained such a strong following in the modern scientific community.

The creation idea leaves many unanswered questions. We should squarely face the mind-boggling possibility that there was no creation of the cosmos; that the universe always has existed and always will exist. To stray away from this insight could lead us in the wrong direction.

Parmenides clearly saw this danger:

"For never shall this be proved: that things that **are not, are**. But do restrain your thought from this path of inquiry, and do not let habit born from much experience compel you along this path, to guide your sightless eye and ringing ear and tongue. But judge by reason the highly contentious disproof that I have spoken."

In other words, don't let your thoughts stray to creation since it implies that existence could evolve from nonexistence, which is impossible. If people had heeded this advice, philosophy and science would have benefited greatly.

We know that the universe appears to expand and that time progresses, which seems to rule out eternal existence. However, the expansion may take place in the cosmological scale beyond space and time, which would allow eternal existence while retaining all the features of the universe. This novel expansion mode not only resolves numerous conceptual problems with the SCM, but it also better agrees with astronomical observations. Most important, the new theory is in accordance with known science (except for a reasonable generalization of GR). Also, as we shall see, the cosmological scale-expansion is a natural consequence of a five-dimensional cosmos with the scale of spacetime as the fifth dimension.

But, even if we were to accept the Big Bang creation at face value, several serious problems would still remain with the SCM. I will mention a few of the most obvious and well-known riddles and show how the SEC theory resolves them.

The Horizon Enigma

SCM problem: When we look out at great, and diametrically opposite, distances in the universe, we find that they look very much the same. The distribution of galaxies is very similar in opposite directions and the cosmic microwave background is almost identical (except for the dipole). This is very difficult to explain in the SCM since light arriving to us from very distant regions would not have had time to reach galaxies located far away in the opposite direction since the Big Bang creation. Therefore, distant regions in one direction have never "communicated" with regions in the opposite direction, so why do they look so similar? This feature must mean that the primordial universe was created with isolated regions that from the very beginning were very similar, which from a scientific point of view is untenable.

This puzzle was "solved" by a speculative invention: *Inflation*. By this scenario, the cosmological expansion was initially fairly slow, allowing different parts of the universe to reach "energy-equilibrium" that would make all regions look the same. At this very early time, everything within the universe was compressed into an extremely dense state, which presumably later expanded much, much faster than the speed of light for a short time. This is called the *inflationary phase*. Then, the expansion slowed again to match what we see today. Expansion faster than the speed of light is excused by saying that we are not dealing with motion *in* space, but that space itself expanded much faster than the speed of light!

The inflation scenario circumvented a major obstacle with the SCM. However, many suspect that inflation merely is an invention purposely designed to save the Big Bang theory.

The SEC theory's solution to the horizon problem: The SEC theory easily resolves the horizon enigma. Eternal existence and unlimited spatial extension allows thermal equalization across arbitrarily vast distances.

The Omega Enigma

SCM problem: Omega refers to the Greek letter Ω, which is used to denote the ratio between the cosmological matter density and a matter density first found by Einstein [Einstein,

1917], the so-called "Critical Density". If omega equals one, the actual mass (energy) density in the universe equals the Critical Density. In the Big Bang scenario, all galaxies recede from each other, creating the observed cosmological redshift. We can estimate what the matter density should have been in the past in order to have resulted in the density we see today, assuming that all galaxies from the beginning were tightly clustered and that the expansion is being slowed by the gravitational pull between them. When we do this, we find that the density in the past must have been very, very close to the Critical Density. As a matter of fact, the initial density must have been almost exactly equal to the Critical Density. If the initial density had been just a tiny fraction higher, gravitation would have halted the expansion, it would have become contraction, and the universe would already have collapsed into a "Big Crunch." With a slightly lower initial density, the expansion would have been faster and would have resulted in a much lower mass density than what we see today. This is quite a coincidence for which there is no scientific explanation. Fortunately for the SCM supporters, the inflation theory took care of this problem, but it created another problem instead by predicting that the cosmological energy density today still should equal the Critical Density. This is far from the case if the cosmological energy density is dominated by visible matter, which means that there must be a lot of unseen Dark Energy throughout the universe.

SEC solution to the Omega problem: The SEC theory resolves the Omega problem in a simply way, since cosmological scale-expansion preserves distances between galaxies and therefore the average mass density, which does not have to equal the Critical Density. Also, there is no missing Dark Energy in the

SEC; observations agree with the model's predictions. What in the SCM appears to be missing Dark Energy is in the SEC, spacetime energy generated by the scale-expansion. We will return to this in chapter 6 when presenting the cosmological vacuum energy-momentum tensor of the SEC.

The Arrow of Time

SCM problem: In the standard physics of the SCM, equations of motion are time-symmetrical; they allow time to run backwards as well as forwards. This is obviously not true in the macroscopic world; we all know that time always runs forward. There is a definite direction of time, or "arrow of time", pointing from the past to the future, but the SCM does not explain why this should be the case.

SEC solution to the arrow of time: The cosmological scale expands with time, which is the cause of the arrow of time. Processes in the SEC are no longer time-symmetrical as evidenced by cosmic drag, which slowly reduces relative velocities and angular momentums for all freely moving objects.

Entropy

SCM problem: Entropy is a measure of how disorganized a system is. In a closed system, which is isolated from its environment, energy exchanged in various processes should always increase the disorder of the system until equilibrium is reached and no further energy exchange is possible. If the cosmos is such a closed system, it should, with time, tend toward increasing disorder, yet it seems that the universe is more organized today than it must have been in its violent Big Bang beginning. Somehow, the superheated primordial gas that existed after the Big Bang has organized into matter, forming galaxies, suns, planets, and people in violation of the demand that the disorder should increase with increasing entropy.

SEC solution to the entropy problem: The SEC is not a closed system in the classical sense since spacetime expands and, in the process, induces energy via the slowing progression of time (see further below). This cosmologically generated energy sustains the universe and keeps the entropy constant. Classical thermodynamics does not take into account the possibility that the scale of time may be changing.

Regions beyond Our Universe

SEC problem: A curious property of the SCM is that the more distant a galaxy, the more rapidly it moves away from us. There is a distance, the Hubble distance, where galaxies recede at the speed of light and the redshift becomes infinite. Regions beyond this distance can never be seen, and for all practical purposes, they do not belong to "our universe." This is a peculiar consequence of the SCM model.

SEC solution: In the SEC, the whole universe is technically within the horizon. In principle, there is no limit to how deep and how far back in time we can see into the universe, but in practice the redshift, with its loss of photon energy, implies a practical limitation. In the future, better telescopes will allow us to see even deeper.

The Age Enigma

SEC problem: The age enigma clearly presents an indisputable problem for the SCM. There are stars in globular clusters in the Milky Way that seem to be older than the 13.8 billion years that currently is the accepted age of the universe. Also, the Hubble Space Telescope (HST) has detected fully developed galaxies with metal-rich stars at distances close to 13 billion light-years. Since these metals must have been created in earlier supernova explosions, the universe must be much older than 13.8 billion years. Even more problematic are observations revealing huge superclusters of galaxies separated by enormous voids; these structures would take several hundred billion years to accumulate if they were to form by gravitational attraction. This is an undeniable, irresolvable challenge to the SCM.

SEC solution: The age problem simply disappears if the there is no cosmological age limit.

HST: Hubble Space Telescope

The Hubble Space Telescope took deep-sky photographs in the visible wavelengths in 1995 and in the infrared wavelengths in 1998 that imaged faint galaxies within about 5 percent of the time to the supposed Big Bang event. The visible images from 1995 show lumpy galaxies that seem to indicate that the galaxies at that time were different from nearby galaxies. However, this was wrong because due to the redshift they were only imaging the young blue stars in the galaxy.

When this same field was imaged in infrared in 1998, the older, redder stars showed up. Their light had been redshifted out of the visible and into the infrared part of the spectrum. Galaxies that initially looked lumpy turned out to be spiral galaxies much like our own Milky Way, which contained many old stars. This implies that these galaxies are far older than the age of the universe according to the SCM. Some theorists have suggested that a galaxy must undergo many hundreds of rotations to form a spiral shape, but our own spiral galaxy is rotating at a rate that only allows a few tens of rotations since the Big Bang. At this rate, those spiral galaxies seen in the deep field would have had time to rotate only two or three turns.

Careful measurement of the distance between galaxies shows that they are regularly spaced, and are not closer together as we look back in time, nearer to the time of the Big Bang. One would expect them to be closer together the farther back we look in time.

The Cosmic Microwave Background

SCM problem: Initially, when the CMB was discovered, it was believed to be the remaining radiation from the Big Bang, thus confirming the SCM. However, later, when more accurate data became available, the CMB was found to be very uniform; in fact, it is *too* uniform. The highly isotropic nature of the cosmic background radiation indicates that the early stages of the universe were almost completely uniform. This raises a problem in explaining the formation of large-scale structures like galaxy clusters, which should give rise to energy variations. The SCM theory predicts larger fluctuations than those we actually see.

The recent discovery of small deviations from smoothness (anisotropies) in the CMB was welcome for the SCM, for it provides possible seeds around which structures might form in the later universe. However, the CMB is still too smooth, and we are far from a quantitative understanding of how the CMB came to be.

SEC solution: In the SEC, the CMB is thermally equalized electromagnetic radiation. Four-dimensional scale-expansion preserves the Planck black-body spectrum without cooling [Masreliez, 1999]. Redshifted electromagnetic radiation from distant regions in the universe has equalized over eons, forming a Planck black-body spectrum—much like radiation energy is equalized in a black-body cavity here on Earth.

Black-body (Planck) Spectrum

When matter is heated, it starts radiating electromagnetic energy. If the temperature is high enough, the radiation becomes visible—the matter starts glowing. The shape of the radiation spectrum is always the same but the location of the peak moves to higher frequencies with increasing temperature. This is the black-body spectrum, or the Planck spectrum, named after its discoverer Max Planck. Historically, this spectrum is of interest since it may be explained if light only comes in discrete quanta. This discovery by Max Planck became the first step toward the subsequent development of Quantum Mechanics.

Black Holes

Black holes would be truly fascinating objects if they existed.

However, black holes do not exist in the SEC.

Initially the black hole idea was suggested by Karl Schwarzschild's exterior solution to Einstein's GR equations, published in 1916, based on the assumption *that the vacuum energy density disappears*. However, in the SEC, this density does not disappear because the scale-expansion induces temporal and spatial vacuum energy that cancel. Therefore, with the scale-expansion the black hole solution to the GR equations no longer exists.

Einstein rejected the black hole idea since it would imply existence of a spacetime singularity with infinite mass density, which defies realistic physics. Recently, consensus has been building about the cores of all galaxies that contain regions with very high mass densities, which are believed to be black holes. But, strangely, these regions usually contain merely a fraction of a percent of the total mass of the galaxy. If black holes existed it would be very difficult to explain why the mass of the central black hole always seems to be a constant small fraction of the total galaxy mass. Why doesn't the black hole keep growing? What can possibly limit its growth to a particular size?

Theoretically, there are no black holes in the SEC, since the energy-momentum tensor in vacuum does not disappear. We find that black holes cannot form. A particle cannot fall inside the event horizon, the "radius of no return" for a black hole. This follows from the GR equations for the SEC theory [Masreliez 2004c].

The Energy-Momentum Tensor

A *tensor* is a mathematical matrix containing quantities called *components*.

The *energy-momentum tensor* is a four-by-four matrix consisting of 16 components, which specifies how energy is distributed in space and time. In an isotropic and homogenous universe (without shear stresses), it reduces to four components, one for time and three for space, out of which the three spatial components are equal. The energy-momentum tensor specifies the energy density in curved spacetime.

Extending this thinking to the cosmos, this tensor also controls the expansion of space in the Big Bang universe. Thus, the cosmological mass density controls the Big Bang expansion.

However, there is an opposite, equally valid, point of view—that the curving of spacetime defines the energy-momentum tensor. This would make *spacetime* rather than *mass density* the primary constituent of the universe, which is the case in the SEC where the energy-momentum tensor does not disappear but contains positive and negative energy induced by the scale-expansion that add to zero. It is the source of the mysterious Dark Energy.

Active Galactic Nuclei

It appears that the velocities of stars observed close to the cores of galaxies are very high, indicating the presence of huge mass accumulations at their cores. Sometimes these galactic nuclei are active (AGNs) and are emitting light and ejecting gas in jets. In the SEC context these mass accumulations are not surprising because stars are continuously falling inward toward the galaxy's core.

The current opinion among astronomers is that most, if not all, galaxies contain black holes at their centers. Typically, the mass of these so-called black holes is only a small fraction of that of the mother galaxy. This suggests that some mechanism must prevent the mass concentration from becoming arbitrarily high.

In principle, this is in agreement with the SEC theory, according to which black holes cannot form. Also, in an eternal universe like the SEC, galaxies cannot grow arbitrarily large. The question becomes: what might be happening instead? Let's speculate.

I have published a preliminary analysis [Masreliez 2004c] that shows that the energy density may become *negative* very close to the event horizon, which would have the effect of cancelling gravitational attraction; it would become repulsion rather than attraction. Furthermore, the analysis also indicates that this conversion to negative energy happens in a very narrow region, very close to the event horizon. We might speculate that matter falling into this region will be strongly repulsed, which might explain the sometimes violent AGN activities. Furthermore, since matter is falling inward along spiralling trajectories, it will tend to converge toward the axis of rotation in ever-narrowing spiral orbits until the centrifugal force no longer can keep it from falling inward. This might explain the narrow jets ejected from the poles of a spinning galaxy core; these jets could be caused by matter falling inward and being repulsed at the poles. The SEC theory could therefore account both for the limited size of the central mass accumulation and for the ejected jets.

Quasars might be the result of two colliding galaxy cores. The massive merging of the concentrated mass accumulations at the core of two galaxies could cause huge conflagrations, with the ejection of an enormous amount of matter combined with intense radiation. This radiation might emanate from deep inside a gravity-well and thus account for at least some of the observed quasar redshifts. This would explain why quasars and their mother galaxies sometimes show different redshifts.

Quasar Distribution

Zhuck et al. analyzed the spatial distribution of quasars based on the distance gauge $d = R_o \ln(1 + z)$, where R_o is in the order of 10^{26} meters [Zhuck et al., 2001]. This distance gauge is identical to the SEC theory's redshift–distance relation if the Hubble distance is about 11 billion light-years. Zhuck et al. concluded that the quasar distribution is *uniform* without any indication of spatial or temporal variation, which supports the SEC theory by showing that no particular age is associated with the existence of quasars. This contradicts the SCM claim that quasars were more prevalent during a certain epoch in the past—the epoch between 1.9 and 3 billion years after the creation. It appears that this claim is based on the erroneous distance gauge of the SCM, and exemplifies how a wrong model could lead to wrong conclusions and to evolutionary speculation.

Pulsar Spin-Downs

Although the cosmic drag effect is quite tiny and very difficult to detect in the planetary motions, it may be detected and directly measured in the spin-down of pulsars.

Pulsars are believed to be strongly magnetized rotating neutron stars, which emit rotating electromagnetic radiation beams much like the beams from lighthouses. These beams are detected by radio telescopes as regular pulse trains with periods indicating the pulsar rotation rates. These pulse trains are extremely stable, but on average, the rate of rotation decreases very slowly with a time-constant that agrees with the SEC model's prediction.

If the spin-down were caused by cosmic drag, we would expect the period to increase exponentially due to loss of angular momentum:

$$p = p_0 \cdot e^{t/T}$$

Here p_0 is the initial rate at some arbitrary time $t=0$. This may also be written:

$$p(dp/dt)^{-1} = T = 10 - 14 \text{ billion years}$$

In one of my papers [Masreliez 2000], I list 25 pulsars with periods ranging from 1.6 milliseconds to 1.0 second together with their corresponding spin-down rates, dp/dt. In spite of that fact that their rates of rotation are varying by a factor 770, 17 of them give values of the Hubble time, T, as given by the relation above, in the range 4 to 25 billion years, corresponding to a factor 6 variation. This strongly suggests that the pulsar spin-downs have a common origin, with cosmic drag being a possible explanation.

The SCM versus the SEC: Concluding Comments

After reviewing the observational, conceptual, and philosophical advantages of the SEC model, we must conclude that the SEC is a superior alternative to the SCM.

Here is a quote from Alice in Wonderland:

"Alice laughed: "There's no use trying," she said; "one can't believe impossible things."

"I daresay you haven't had much practice," said the Queen. "When I was younger, I always did it for half an hour a day. Why, sometimes I've believed as many as six impossible things before breakfast."

The SCM expects people to believe more than six impossible things. Although we may have composed beautiful theories and developed elaborate mathematical tools in support of the SCM, it does not mean that the physical world complies with these theories and follows these rules. Occasionally science, as currently understood, might lead to the wrong conclusions.

It is said that Alexander undid the Gordian knot by parting it with his sword. Perhaps scholars of his time cried foul, but he solved the problem, even if his method was unexpected. The SEC theory with its semi-discrete scale-expansion, which explains the progression of time, could be the sword that cuts the Gordian knot of cosmology, resolving a number of problems.

I round out this chapter with an illustrative story shared with me by my good friend Dan Odell.

A man goes to the department store to buy himself a new suit. When looking around, he finds a very attractive one and asks the salesperson about it. It turns out that it is on sale at half price. Unfortunately, it is three sizes too large but is the only one left. Seeing that the customer really likes it, the salesperson suggests that the pants might fit if he just could grab hold of the legs and pull them up a little on each side. And they would not fall down if he just could hold his arms close to the body. Also, the jacket would look good at least in a frontal view if he smoothed it out before locking both arms to the sides, and this would also prevent the jacket sleeves from falling down. Thus, with a little effort he could make the suit fit quite well. Convinced (since he really wants the suit and since the price is right), the man buys it and decides to put it on right away. When leaving the store, he naturally finds it a bit difficult to walk.

A couple comes by and sees the man hobbling along. The woman says to her husband, "Did you see that poor handicapped guy?"

"Yes," answers the husband. "Poor guy, but his suit looks great."

Like the suit, the SCM might look great, but it simply does not fit.

CHAPTER 5

Observational Evidence within our Solar System

IT IS NOT UNCOMMON FOR ENGINEERS TO ACCEPT THE
REALITY OF PHENOMENA THAT ARE NOT YET UNDERSTOOD,
AS IT IS VERY COMMON FOR PHYSICISTS TO DISBELIEVE
THE REALITY OF PHENOMENA THAT SEEM TO CONTRADICT
CONTEMPORARY BELIEFS OF PHYSICS.

—H. BAUER

Engineers want to make things work regardless of how it is achieved, while physicists do not want to acknowledge that things work unless they can explain how.

In this chapter, I address the observational support of the SEC theory found within our solar system, beginning with the so-called Pioneer Anomaly, and show that this puzzle is closely related to two other riddles: (1) the recently detected disagreement between optical observations of the inner planets and their computed positions as given by their ephemerides and (2) the erroneous conclusion that the Moon must have been in contact with the Earth some 1.5 billion years ago.

The SEC theory explains these seemingly unrelated problems as being caused by misunderstanding the true nature of spacetime within our solar system. This chapter goes deeper into the details of the SEC theory and demands a bit more from the reader. However, I think it should be of interest since it demonstrates how an inconspicuous error in perception may have caused us to misinterpret our world.

The material in this section is presented in [Masreliez 2005b].

The Pioneer Anomaly

On March 2, 1972, Pioneer 10 was launched on an Atlas/Centaur rocket from Cape Canaveral. Pioneer 10 was the first space probe sent to the outer planets. After surveying Jupiter on December 4, 1973, it continued outward in the plane of the ecliptic and became our first spacecraft to leave the planetary part of the solar system when it passed beyond the orbit of Pluto in 1983. It was last heard from on January 22, 2003, at a distance of approximately 82 astronomical units from the Sun (AU; one AU is the distance between the Earth and the Sun). At that point, the roundtrip time for the radio signal exceeded 20 hours!

A microwave carrier wave signal transmitted to Pioneer 10 was returned with preserved coherence (phase relationship) and was monitored until June 1998. Analysis of the signal at the Jet Propulsion Laboratory (JPL) has yielded new and unique information as reported by Anderson et al. [Anderson et al. 2002].

The outward motion of Pioneer 10 was estimated by two independent methods.

First, the frequency shift between a signal transmitted uplink from the Earth and returned downlink by Pioneer 10 was compared to a reference signal. This frequency shift is (mainly) a Doppler shift due to the outward motion of Pioneer 10 and therefore estimates the velocity.

Second, the distance to the space probe was measured by modulating the phase of the transmitted signal by short pulses at regular intervals, which later were detected in the received downlink signal. The time difference between the transmitted and returned pulses was used to estimate the distance, since the electromagnetic signal moves at the speed of light (except for tiny gravitational corrections). This range information was then used as inputs to ephemeris modeling programs at JPL that estimate the velocity of Pioneer 10 and from this velocity derive an estimated Doppler-shifted frequency that should be received at the downlink. This estimated Doppler shift was then compared to the directly measured Doppler shift.

However, a small difference between these two methods of estimating the velocity became apparent during the 1980s. It appears that the space probe is subjected to a tiny, constant acceleration toward the Sun of unknown origin that depends neither on time nor distance. No physical explanation has yet been found. Although claims recently have been made that the acceleration actually seemed to decrease over a longer time span the fact that the same acceleration anomaly has been detected with several different spacecrafts indicate that origin of the anomaly is of systematic origin [Turyshev et al, 2011].

The Pioneer acceleration anomaly is four orders of magnitude larger than the SEC theory's cosmic velocity drag. However, as we shall see, it may still be explained by the SEC theory.

Since the received frequency is slightly higher than the estimated, the difference might be caused by an acceleration of unknown origin toward the Sun (in the line of sight) of $(8.74 \pm 1.33) \cdot 10^{-8}$ cm/sec^2. Recent reprocessing of the data over a longer time interval (23 years) published in 2011 shows that the anomalous acceleration seems to have decreased with time and stabilized at around $7.5 \cdot 10^{-8}$ cm/sec^2 [Turyshev et al, 2011]. Furthermore, the same acceleration has been detected for Pioneer 11 and for the Galileo and Ulysses spacecrafts. This indicates that we are dealing with a new phenomenon, possibly of cosmological origin. Over the years, many possible explanations for the Pioneer anomaly have been suggested, but so far, none of them has explained it. Most of these suggestions investigate different mechanisms that might cause the observed small acceleration, such as gas leaks from jets controlling the attitude of the spacecraft. However, a few contributors have noted that the Pioneer anomaly does not necessarily imply acceleration; the detected discrepancy occurs between two ways of deriving a certain frequency shift, and this does not necessarily imply a velocity difference. People have also noted that the measured acceleration seems to be close to c/T (the speed of light divided by the Hubble time), suggesting some kind of cosmological connection. If the acceleration equals c/T the more recent data corresponds to T=13.3 billion years in good agreement with estimates for the Hubble time based on other methods. The paper by Anderson et al. summarizes these efforts.

Figure 11 illustrates the Pioneer data tracking.

The Pioneer Anomaly
Two ways of estimating velocity.

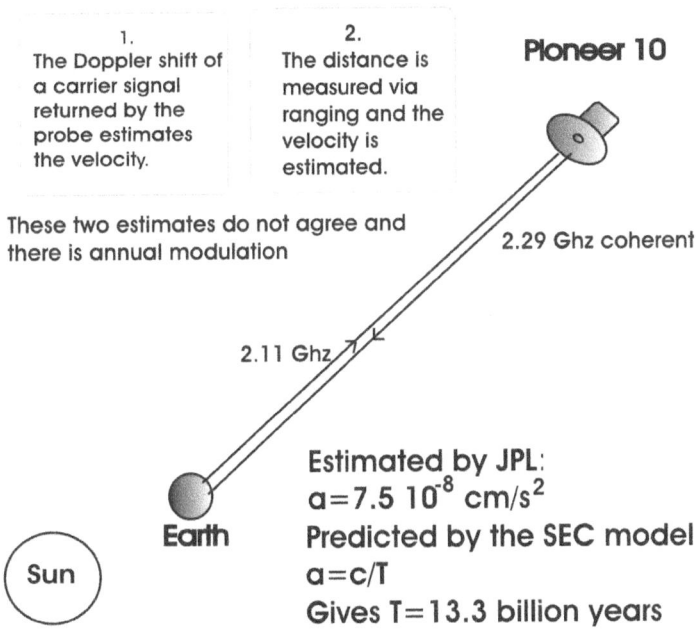

1.
The Doppler shift of a carrier signal returned by the probe estimates the velocity.

2.
The distance is measured via ranging and the velocity is estimated.

Pioneer 10

These two estimates do not agree and there is annual modulation

2.29 Ghz coherent

2.11 Ghz

Earth

Sun

Estimated by JPL:
$a = 7.5 \ 10^{-8} \ cm/s^2$
Predicted by the SEC model
$a = c/T$
Gives T = 13.3 billion years

Figure 11: The Pioneer Anomaly

The Pioneer anomaly is explained by the SEC theory, which predicts that the discrepancy actually is c/T, but before I show you how, I will present another related puzzle.

The Planetary Ephemerides

The ephemerides tabulate planetary positions on the sky as seen from the Earth. The construction of planetary ephemerides probably is the most ancient task of astronomy beginning with early observations of so-called wandering stars, that is, the planets. Johannes Kepler made a very significant contribution based on observations collected by the Danish astronomer Tycho Brahe. He determined that the planetary orbits are elliptical [Kepler, 1697]. Isaac Newton brilliantly demonstrated that these elliptical orbits are a consequence of his laws of motion and gravitation. It became clear that the planets might move freely around the Sun without slowing down, with orbits determined by gravitational attraction.

The new Copernican worldview dramatically changed the techniques used for of ephemeris construction. With the shape of the orbits now believed to be known, they could be fitted to the observations merely by adjusting a few parameters of an ellipse. The method used today is still based on Newton's laws of motion and gravitation with relativistic corrections. It also takes into account the combined gravitational influences from different planets and asteroids. The modern ephemerides generated by JPL are based on numerical integration where the motions of all the planets are computed simultaneously, using repeated iterations in order to arrive at the best fit of the observational data to the theoretical orbits. This is a demanding computational task using techniques developed and refined over centuries [Standish and Williams, 1990].

The parameters determined by this approach of fitting the observations to the theoretical orbits also include *fitting their time-base*. In other words, the times at various locations in the orbit are determined so that the orbits fit the mathematical model. This is crucial; since an accurate temporal reference was missing in the past the planetary motions were actually used as a clock with a rate determined by fitting the observations to orbits determined by Newton's laws. Today JPL uses the same approach; the assumed mathematical model together with the observations determines the time-base.

Modern Ephemerides

Modern ephemerides are used by NASA and other agencies in the space program. However, the planetary observations will with time drift away from their by JPL-estimated ephemerides. Although unexplained, these discrepancies have now become an accepted nuisance and are in practice handled by periodically updating the ephemerides based on more recent observational data, while repeatedly discarding older observations that are considered unreliable.

Time in Astronomy

The question of a time-base has always been of central importance for astronomy. Traditionally, positions of the planets were recorded by noting the year, the date, and the time of day of the observations. Hence early on, the clock used in astronomy was the rotating Earth. This solar time, or Universal Time (UT) as it is now called, was used from the beginning of modern astronomy until the middle of the twentieth century when it became clear that UT was no longer accurate enough for astrometry, because the rotation of the Earth fluctuates due to influences like ocean tides, winds, inner magma flow, and so on. The motion of the Earth around the Sun became a more accurate temporal standard, which could be derived from fitting the observations to Newtonian orbits as predicted by Newton's laws of motion and gravitation. However, this time-base was difficult to use in practice. For a brief time in the middle of the twentieth century, Ephemeris Time (ET) was the temporal standard in astronomy until it was replaced by the more easily accessible Atomic Time (AT) in 1955.

UT, ET, and AT

Universal Time (UT) is based on the rotation of the Earth, Ephemeris Time (ET) on its motion around the Sun, and Atomic Time (AT) on submicroscopic oscillation of atoms or molecules. Since 1955, AT is in general use in science, but the planetary motions are still computed by JPL using a form of ET called T_{eph}, which is found by fitting the orbits to observations.

Meanwhile, the method of fitting the ephemerides to observations was steadily being refined by adding relativistic corrections and by taking into account gravitational influences from the other planets and from several of the largest asteroids. Also, computer programs that automatically fit the observations to the Newtonian orbits from which ET could be determined were developed by JPL.

After the introduction of AT in 1955, the continued use of a fitted time-base was challenged by the suggestion that AT should replace ET. However, JPL rejected this suggestion. There might have been two main reasons for this decision: first, the classical, proven approach was still in use and a large investment had already been made in developing techniques that simultaneously and automatically fit the time-base to the observations. The use of AT would require revision of this approach and would obsolete computer programs already developed at great cost [Standish, 1998]. Second, the fit of planetary observations to Newtonian orbits proved to be excellent, which seemingly confirmed the equivalence of ET and AT. If this were the case, nothing would be gained by revising the established methodology.

In the 1970s, a new program was initiated by which distances between the Earth and the planets Mercury, Venus, and Mars were measured using radar ranging. If a radar pulse is sent in the direction of one of these nearby planets, the distance it travels can be determined quite accurately from the radar echo. The accuracies of these measurements far exceed those of the optical observations, but with one caveat: optical observations measure planetary positions relative to the stellar background while there is no such external reference with the ranging measurements. However, this obstacle may be overcome by combining range measurements with optical and Very Long Baseline Interferometry (VLBI) measurements. Modern ephemerides published by JPL rely heavily on range data and VLBI measurements. The ephemeris time-base, which JPL now calls T_{eph}, is still fitted to the observations, assuming Newtonian orbits. (JPL uses the notation T_{eph} rather than ET since these two time-bases are slightly different.) T_{eph} is then adjusted to AT as closely as possible with the assumption that it is proportional to AT [Standish, 1998].

However, lately disturbing and unexplainable discrepancies have surfaced. Optical observations drift away from the computed ephemerides, which have to be updated at regular intervals by adding new observational data while discarding older data that are deemed unreliable. The general belief is that there might be some kind of modeling shortcoming or that optical observations are afflicted by some kind of systematic error, since it is well-known that the ranging data is superior.

Today the drift of the optical observations relative to the ephemerides is an irrefutable but unexplainable fact. Recently several independent investigators have reported discrepancies between the optical observations and the planetary ephemerides. Some research [Yao and Smith 1988, 1991, 1993; Krasinsky et al. 1993; Seidelman et al. 1985, 1986; Kolesnik, 1995, 1996; Poppe et. al. 1999, Kolesnik & Masreliez, 2004] indicate that residuals of right ascensions of the Sun show nearly a one arc-second per century negative drift before 1960 and an equivalent positive drift after that date. Yuri Kolesnik reports on positive drifts of the planets relative to their ephemerides based on optical observations covering thirty years with atomic time. He uses data from many observatories around the world, which all independently detect these planetary drifts [Kolesnik, 1995, 1996]

Explaining the Planetary Observational Discrepancies

All of this is, of course, very intriguing, and one wonders what might cause these observed planetary drifts. In order to understand what might be going on, we have to dig a bit deeper.

According to the SEC theory, spacetime is curved, not just cosmologically but even locally in the solar system. Here, the term *curved* means that the cosmological expansion influences the geometry of the 4D spacetime, which causes new phenomena. For example, we saw that the SEC model implies cosmic drag and that the planetary orbits no longer are Newtonian but follow spiral trajectories toward the Sun.

But JPL has shown that the orbits are almost perfectly Newtonian (which with relativistic adjustments sometimes are called post-Newtonian, but I will simply call them Newtonian). How can this be possible? The excellent agreement between observations and the computed ephemerides found by JPL seems irrefutable and appears to rule out the SEC theory as well as any problem with the ephemeris construction process.

Unfortunately, this is not the case.

Although spacetime by the SEC theory is curved, according to GR a locally flat approximation (Minkowskian spacetime; see sidebar) always exists for any curved spacetime. This may be compared to a flat tangent plane that locally approximates a curved surface of, for example, a sphere. If the assumed coordinates are Minkowskian, Einstein's equations predict Newtonian orbits. Therefore, if we use the coordinates of a Minkowskian tangent spacetime instead of the curved coordinates of the SEC, *the planetary orbits will automatically become Newtonian!* In other words, Newton's laws hold true with the "right" choice of coordinates. And, with this choice of coordinates, discrepancies from Newton's laws cannot be detected.

Minkowskian Spacetime

This is also called "flat spacetime," in which all metrics are constant. It is named after the mathematician Hermann Minkowski who expressed SR as a four-dimensional theory. Minkowski had also been one of Einstein's teachers and had considered him "a lazy dog," but later admitted that maybe his former student wasn't that lazy after all.

It can be shown that the Minkowskian tangential coordinates fit the curved SEC coordinates with a fractional error of about 10^{-28} in the inner solar system, which is totally negligible. The corresponding distance error between the Earth and the Sun would be about 10^{-17} meters! Also, the law of gravitation in the SEC differs from Newton's law, see Appendix III. Therefore, when fitting the planetary ranging observations to Newtonian orbits, the computer program will *automatically select the flat Minkowskian spacetime* rather than the SEC curved spacetime, and the observations will (almost) perfectly fit the orbits. In other words, assuming that the orbits are Newtonian will with the selected coordinates guarantee that that the observations fit!

However, the fact that the ranging observations fit Newtonian orbits does not mean that the AT and T_{eph} time scales are proportional.

Arguing that AT must be equal to T_{eph} since the orbits are Newtonian amounts to circular reasoning, since the presumption that the orbits are Newtonian automatically selects coordinates for which this is true. With the Minkowskian coordinates, the planetary orbits are (almost) perfectly Newtonian. But what is wrong with this? Can't we use these Minkowskian coordinates instead of the SEC coordinates? Yes, we could, and we actually did this before introducing AT. In fact, we might not have detected the difference at all but for the existence of AT. The problem arises mainly because the time-bases for the two coordinate representations do not agree.

According to the SEC model, T_{eph} accelerates relative to AT.

Therefore, the observational discrepancies with the planetary ephemerides may be due to the inadvertent use of two different coordinate systems; JPL uses T_{eph} while the optical observations use AT. Eventually this temporal acceleration discrepancy will become glaringly apparent. A few studies have already discovered it, for example the one by Oesterwinter and Cohen who constructed planetary and lunar ephemerides with AT as the time-base rather than fitting the time-base to the observations [Oesterwinter and Cohen, 1972]. They found that ET drifts positive relative to AT. Also, using AT two teams, one American and one Russian, independently found that the planets accelerate based on ranging data [Reasenberg and Shapiro, 1978; Krasinsky et al., 1986].

If these planetary accelerations really exist, one might rightly wonder why they haven't already been discovered. The explanation might partly be that the traditional approach of fitting the ephemerides described above hides the accelerations. Since the ephemerides are fitted mainly by using ranging data, the secular (angular) drift in relation to the stellar background is not detected. However, at the present time some 40 years after the inception of the ranging program, the planetary accelerations should become noticeable—at least for Mercury, for which the drift is largest. On the other hand, optical observations, for which over 50 years of observational data are available based on AT, detect secular acceleration of the planets relative to the stellar background.

However, there is another possible explanation to why the drifts have not been acknowledged.

A MAN RECEIVES ONLY WHAT HE IS READY TO RECEIVE...
THE PHENOMENON OR FACT THAT CANNOT IN ANY WISE BE
LINKED WITH THE REST OF WHAT HE HAS OBSERVED, HE DOES
NOT OBSERVE.

—H. D. THOREAU

In the belief that the Newtonian model is absolutely correct, the answer to these puzzling discrepancies is being searched for elsewhere, but in the wrong places. The possibility that the orbits do not follow Newton's laws is simply unthinkable for the experts. However, in the SEC where there is cosmic drag Newton's laws no longer hold. Consequently the coordinates that make Newton's law valid are not the right cosmological coordinates. In particular, the time-base obtained from fitting the ephemerides does not agree with atomic time. The relationship between ET and UT in view of the SEC theory is discussed in my Pioneer Anomaly paper [Masreliez, 2005b]. Furthermore, Appendix III discusses the difference between the flat Minkowskian coordinates and the SEC curved coordinates.

Explaining the Pioneer Anomaly

The Pioneer anomaly might also be caused by using different coordinate systems. We saw that it arises from two different methods of estimating the velocity. The first directly measures the Doppler shift between the uplink and downlink signals. This estimate is *based on AT.* Atomic clocks stabilize the frequency and are used in all other timing measurements.

On the other hand, although the distance is measured by ranging using AT, the velocity of the space probe is estimated using the same type of ephemeris tracking program as is used in computing the planetary ephemerides. The time-base here is T_{eph}. This partly explains the Pioneer anomaly, but there also is an additional contribution from a small difference in the radial coordinate due to spacetime curvature, see further below. The engineers and scientists working in the Pioneer program should realize that they have done a superb job and that there is nothing wrong with either the Pioneer 10 space probe or with their data processing. The problem should instead be blamed on the ephemeris estimation approach. The Pioneer anomaly would have disappeared if the signal frequency had been transmitted and measured relative the time-base T_{eph} rather than AT (which would be quite difficult to do in practice). Alternatively, it would also have disappeared if the mathematical model for the orbits had been taking into account cosmic drag. Thus, the Pioneer Anomaly directly confirms cosmic drag, but it has been ignored by people, who simply cannot fathom the possible existence of a gross modeling error.

The SEC theory predicts that the Pioneer acceleration anomaly should equal c/T, where c is the speed of light and T is the Hubble time. This agrees excellently with the measured acceleration $(8.74 \pm 1.33)10^{-8}$ cm/sec^2, which would correspond to a Hubble time in the range 10.0 to 13.5 billion years [Masreliez, 2005b]. This agreement with the SEC theory is remarkable and should be taken as yet another confirmation of the theory. This analysis is presented in Appendix III.

The attention paid to the Pioneer anomaly is commendable since it is our first indication that we might have misunderstood the nature of space and time. However, resolution of the anomaly is prevented by the reluctance to squarely face the possibility that there is a modeling error. Instead, NASA is soliciting additional funding for further studies.

A Moon Mystery

There is tidal action between the Earth and the Moon that is believed to slow down the Earth's rotation rate. Preserving the angular momentum in the Earth–Moon system (in the standard model) demands that angular momentum lost by the slowing rotation of the Earth is picked up by the motion of the Moon. This should cause the orbit of the Moon to expand; the Moon should be moving away from the Earth. It is estimated that the Moon presently recedes by about 3.8 cm/year.

However, at this recession rate, the Moon should have been in contact with the Earth about 1.5 billion years ago. And, since the Moon was closer to Earth in the past, the tidal action should then have been even greater in the past. But this does not agree with collected lunar rocks, which are at least 3.5 billion years old. There is no indication that the Earth–Moon system has changed during the last 3.5 billion years. This is a puzzling enigma.

The choice of a local Minkowskian coordinate system used by JPL rather than the SEC coordinate system would also explain why the Moon's recession rate has been overestimated. One can show that the relationship between the radial recession velocity, v_m, measured with the Minkowskian coordinates used by JPL; and the recession velocity, v_s, measured with the SEC coordinates, is given by

$$v_m = v_s + r/T; \qquad r = \text{radial distance}$$

The average radial distance to the Moon is 3.83×10^{10} cm, and with $T = 14$ billion years, we get $r/T = 2.7$ cm/year and with T=10 billion years 3.8 cm/Year. Since v_m is estimated at 3.8 cm/year, this means that the true velocity,

v_s, instead might be less than 1 cm/year and that the *Moon very well could have been formed some 4.5 billion years ago at the same time as the Earth*. Hence, the mysterious origin of the Moon may be explained by the SEC model. This is further analyzed in Appendix III.

Why Does Universal Time Seem to Slow Down?

In the past, the rotation of the Earth as given by UT was the time-base in astronomy. However, it was soon recognized that UT appears to slow down relative to ET, because the length of the year measured in UT is steadily decreasing; measured in solar time the year is getting shorter because there are fewer days in a year. This conclusion was based on the presumption that Newton's laws of motion are absolutely correct and, consequently, that the time it takes for the Earth to make its path around the Sun remains the same. This would imply that the observed decreasing number of days per year is due to a slowing rotation of the Earth, which still is the general consensus in astronomy.

However, by the SEC theory this is no longer true since the year gradually becomes shorter when the Earth spirals toward the Sun. Therefore, the number of days it takes for the Earth to circumnavigate the Sun decreases with time. The slowing UT could therefore be caused by the Earth slowly spiralling toward the Sun, just like stars spiral toward galactic centers, combined with a slowing rotation of the Earth due to loss of angular momentum in the SEC. This is analyzed in Appendix II.

Fossil Coral Evidence

Evidence initially presented by John Wells supports the proposition that the number of days in the year has changed over time [John Wells, 1963]. The observational material used for this data consists of fossil corals, brachiopods, and bivalves from the Phanerozoic period and stromatolites and tidal deposits from the Proterozoic period. The growth characteristics of these organisms change daily and also with the season of the year, which makes it possible to deduce the number of days in the year during prehistoric times much like tree rings record the age of a tree. With the assumption that the length of the year remains constant, one can conclude that the *apparent* length of day (LOD) was considerably shorter in the past and that the estimated LOD steadily is growing longer *at an accelerating rate.*

Thus, the evidence shows that the rate of change was slower in the distant past than it is today.

This came as a great surprise, since this finding makes it very difficult to explain the change by a tidal slowing of the Earth's rotation. Since the Moon is currently believed to be receding at the rate of 3.8 cm/year due to transfer of angular momentum from the rotating Earth, it must have been much closer to the Earth in the past. This means that the tidal action should have been greater in the past and therefore that the LOD should have increased at a *faster* rather than slower pace in the past.

However, this is the opposite of what the coral evidence indicates.

The coral data actually reflect the number of days in a year rather than the length of the day. Therefore, a greater number of days in the year may also mean that the year was longer in the past. By the SEC theory, the Earth spirals closer to the Sun at an exponentially increasing rate. This explains not only why the number of days in a year was greater in the past but also why the rate is changing faster today than it did a long time ago. Thus, the coral evidence further supports the SEC model.

The Cosmological Importance of the Pioneer Anomaly

I have elaborated on the Pioneer Anomaly and the planetary drifts because they confirm the SEC theory by local observations here within our solar system. It reveals an aspect of spacetime that is well-known by people familiar with GR but which to others might appear almost magical. Not many recognize that the planetary orbits simultaneously may be both Newtonian and non-Newtonian, depending on the choice of coordinates. It is a matter of perspective; it is a consequence of a most basic feature of GR, which allows us to freely select our coordinates. Unfortunately, this also implies that what we observe might be interpreted differently depending on our choice of coordinates.

This becomes even more important in our perception of the universe. By the SCM's choice of coordinates, the universe had a beginning 10-14 billion years ago, but with the SEC model's choice there never was a beginning. This is a stunning difference in interpretation depending on the perspective we choose, and GR does not tell us which is right! Fortunately, the modern temporal standard in astronomy and physics, which is atomic time, differs from the Newtonian ephemeris time and clearly is in agreement with the SEC model. This may be directly confirmed from observations in the solar system. The Pioneer Anomaly and the drifting planets may be our first clear signal that something is seriously wrong with our current understanding.

But, as always is the case with revolutionary new discoveries that challenge current widely held beliefs, the SEC theory is being ignored by the experts, partly because it is new and unfamiliar, partly because it obsoletes earlier work

It has been said that the major contribution of any new hypothesis is its ability to predict features of the universe not yet well understood or recognized and to do this in a way that may be falsified. The locally curved spacetime geometry with its cosmic drag is such a hypothesis that may be falsified by measurements within the solar system. Although it might be difficult to accept this possibility, presently the will to even consider it seems to be generally lacking.

CHAPTER 6

New Physics of the SEC Model

THE ALTAR CLOTH OF ONE EON IS THE DOORMAT
OF THE NEXT.

—MARK TWAIN

I hope to have convincingly demonstrated the scale-
expansion cosmos model's elegant resolutions of a great
many observational discrepancies both at cosmological
distances and here in our solar system, and its explanation
of several cosmological enigmas. But these advantages
come at a price: they imply that accepted epistemology
will have to be revised, starting with Newton's first law of
motion. This will be a main obstacle for the SEC theory's
general acceptance by the scientific community; but

regardless of this, if the theory turns out to have merit, it should eventually prevail. Furthermore, I have not yet addressed a number of implications that scale-equivalence and a dynamic spacetime scale would have—implications that would not only alter our perception of the world but also spur technological advancement that eventually could take us to the stars.

But before addressing these fascinating possibilities, let's take a look at some aspects of the "new physics."

Timeless Existence

Let me again paraphrase Parmenides, who I think was far ahead of his time:

> *The fact that the universe exists rules out nonexistence. It is inconceivable, and is a logical contradiction, that something that now exists can have originated in nonexistence.*

Let us fully accept this obvious but perhaps somewhat disturbing fact with all its implications.

Throughout human development, our perception of the universe has always evolved. Our place in the cosmos has changed from a flat Earth with an outer edge, to a fixed orb at the center of the universe, to a planet in the solar system, to a solar system in the Milky Way galaxy, and to one out of hundreds of billions of galaxies in world that began some 14 billion years ago. *And now this confinement in time also disappears; no longer is there a beginning or end of time.*

Eternal existence implies something extremely important, which I will call the Rule of Timeless Existence (RTE):

All processes in the universe participate in eternal cycles of timeless existence.

This must be true if the universe is to exist eternally without aging. The universe is the ultimate recycling "organism." The RTE has important scientific implications since cosmological processes that are not self-sustaining must be ruled out. We have already seen one example of this; galaxies maintain their shapes and continue existing over very long periods. Matter in a galaxy continually moves toward the core to be ejected and eventually recaptured by its mother galaxy, or by another galaxy, in an eternal recycling flow. But, the RTE may seem unfamiliar and confusing; for example, how can time progress in a universe that never changes?

The Progression of Time

A possible solution to the puzzle of the progression of time is one of the most important and (to me) satisfying aspects of the new theory.

We have always believed that time progresses at the same continuous pace so that each second is as long as the previous second, but this might not be the case; the duration of a second could change together with the scale-expansion. Since the temporal metric expands together with the spatial metrics, the duration of time intervals like a second increases. However, as I argued earlier, it is inconceivable that the duration of a second can change *continuously* relative to itself. The progression of time relates an earlier time interval to a later time interval via relative scale adjustment, and therefore, the progression of time must occur in discrete steps. This also is a technical requirement of GR, otherwise different epochs would not be scale-equivalent. The scale-expansion process may also be modeled by adding a fifth scale-dimension in GR in addition to the ordinary four spacetime dimensions.

We owe (or should I say blame) the currently accepted way of dealing with motion to Newton and Leibniz, who both claimed they had invented differential calculus. In modeling motion, we use a mathematical trick by which we treat motion as a limit of ever-shortening length and time increments. In mathematical terms, we form the *time derivative*.

Since its invention, differential calculus has become the workhorse of science.

Our indiscriminate use of differential calculus and differential geometry might be responsible for the current crisis in physics and cosmology.

It appears that Zeno with his paradoxes had a point: we have to be very careful when applying differential methods, since *physical processes may exist that cannot be modeled by differential calculus.* We already suspect that this might be the case in quantum mechanics, but we have not yet fully grasped all of its implications. GR is based on differential geometry, which is limited to continuous coordinates. In a continuous manifold locations in space and time change smoothly. No matter how close two points are chosen, there are points that are even closer, which makes the differential limiting process work. Ironically, although the 4D space-time manifold modeled by GR may be continuous, the progression of time that takes place via the increasing scale could be discontinuous; it might be a stepwise process.

Processes might *not* be continuous at the atomic-particle level for an excellent reason. By the SEC theory, the universe expands by increasing the scale of spacetime, and we saw that this must be a stepwise process. After each tiny scale-expansion increment, all four metrics have expanded by the same tiny scale-factor, but the universe always still remains the same relative to an inhabitant, since universes of different scales are equivalent. A (very simplified) representation of the cosmological expansion cycle is illustrated in figure 12.

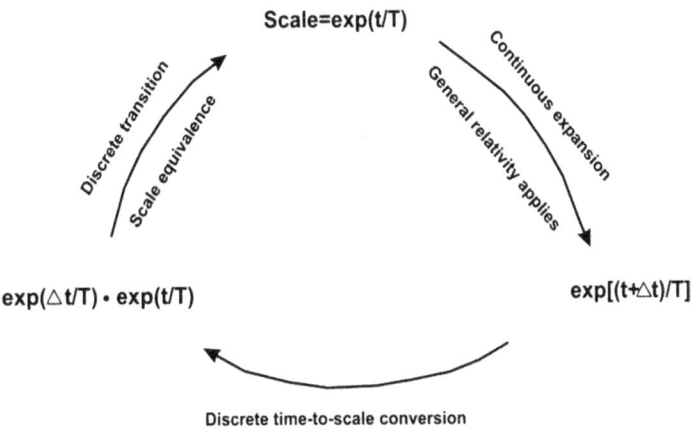

Figure 12: The DIST scale expansion cycle

The cycle in the figure suggests that stepwise scale-expansion that preserves symmetry between the four metrics is the essence of the progression of time.

During the continuous part of the short SEC expansion cycle, the metrics expand. After a short time interval Δt, the scale has increased by the factor $\exp(\Delta t/T)$. Due to scale-equivalence, the universe is at this point in the cycle equivalent to what it was at the beginning of the cycle. The universe "steps into" the new scale by incrementally decreasing the pace of proper time, and the cycle returns to its beginning. This last step can be incorporated in GR by changing the reference increment $ds => ds \times \exp(\Delta t/T)$. This process extends GR to also cover discrete scale transformation, which retains the original line-element. In other words, it extends GR to handle the additional scale dimension. The possibility that the pace of proper time might change with the cosmological expansion is not covered by standard GR.

As I already mentioned you may think of the scale-expansion process as being similar to a child growing out of her clothes. Old clothes, when outgrown, are replaced by new ones on a regular basis. Similarly, the universe changes the pace of proper time incrementally to "fit" the expanding space. The cosmos remains the same, and time always appears to progress at the same pace as experienced by us as inhabitants. The fact that the scale of spacetime changes all the time has remained hidden. It is understandable if you have problems with this novel expansion mode, since it is beyond of common experience as well as known physics. But please remember the guideline:

The cosmos is scale invariant. Therefore, the cosmological expansion could be 4D scale-expansion. We should be able to model this expansion mode mathematically.

My attempt of doing this using GR (since this theory is the best we have) may perhaps not be the optimal choice, but the point is that *scale-expansion should be possible*. An improved mathematical representation will likely be found in the future, possibly based on the five-dimensional extension of GR originally suggested by Theodor Kaluza.

The SEC cycle shown in the figure is of crucial importance. It is a new kind of physical process that will be denoted Dynamic Incremental Scale Transition (DIST). It will be shown that this process not only may model the SEC expansion, but also might explain the phenomenon of inertia. This is the subject of Chapter 7.

Dynamic Incremental Scale Transition (DIST)

Key to appreciating the SEC model is to become comfortable with how an inhabitant who participates in the scale-expansion would experience the world. If you are familiar with the methods currently used in physics, you might perhaps wonder if this new scale-expansion process really is "allowed."

Science is a game played with certain rules, and the new DIST process does not yet belong among these rules. But we must remember that the rules of science were laid down based on previous knowledge. We are continually expanding our knowledge base and revising these rules. The DIST process might be such a revision. I think it is a nice revision because it adds another dimension but does not invalidate GR. The DIST process allows us to have our cake and eat it too.

The incremental scale might change not only for motion in time but also for motion in space. Generally, it may be described by the loop depicted in figure 13, which is a generalization of the SEC expansion loop.

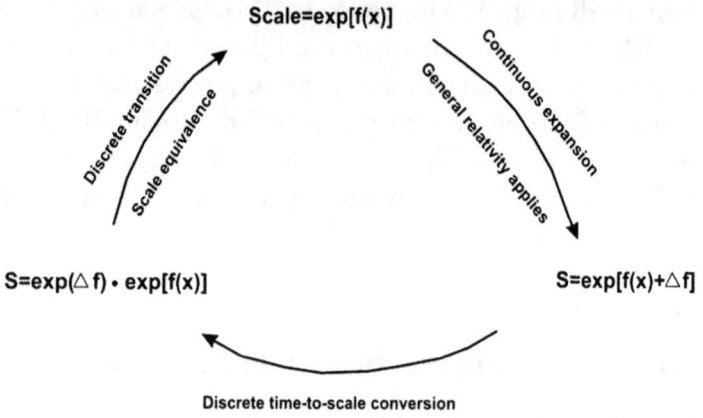

Figure 13: General scale transition cycle

Here, $f(x)$ signifies a function of possibly all four coordinates of spacetime. By the DIST process, the scale of spacetime may change incrementally while all coordinate values remain the same. Therefore, the DIST represents a new mode of motion that takes place "beyond" the 4D manifold of space and time.

It makes use of scale-equivalence, which might be the most fundamental of all symmetries. It is a process that does not change the energy-momentum tensor; it does not "cost" anything, so it can take place without energy loss. It is therefore not surprising that the universe takes advantage of this process in its cosmological expansion mode. And it should not be surprising if this process is also used elsewhere, for example even in ordinary motion in space. In the following, we will see how DIST makes the connection between general relativity and quantum theory and how it explains the origin of the inertial force.

Note that relative to a co-expanding observer the DIST process becomes *cyclic* in nature since the four-dimensional geometry returns to its starting point by the end of each cycle. This explains the eternal aspect of the progression of time that takes place in a "fifth dimension" beyond the four spacetime dimensions. This also explains why the progress of time always has been enigmatic in the past; it cannot be explained mathematically as motion in space or time.

However, we must keep in mind that we are trying to model the dynamic scale process by extending the applicability of known physics. At first this might appear questionable, but it is possible that the DIST process is more fundamental than the traditional continuous processes we are used to. We must acknowledge that continuous processes are achieved by visualizing increments in time and space as being arbitrarily small, which we now know is impossible due to quantum theory and the wave aspect of particles. Therefore, we should not expect that all aspects of the world might be modelled by continuous processes. Continuity may apply for motion in space and time but perhaps not for motion in scale. In other words, our extensive use of differential methods, which have served us so well in the past, may have outlasted their applicability.

The DIST process introduces an additional parameter, a dimension, beyond the four spacetime dimensions.

Universal Perpetual Motion

The RTE seemingly violates known laws of physics, since it appears that the world eventually should run out of energy, ending up in the "heat death." But the SEC universe is different from our common perception because in the SEC the pace of proper time slows down incrementally via the increasing temporal metric.

Heat Death

The perhaps most unattractive feature of the Big Bang theory is the prediction that space will expand forever with forever-decreasing energy. Eventually all the energy in the stars will be depleted, they will stop shining, and the universe will die a dark and cold "heat death." The SEC theory, on the other hand, concludes that there is no heat death. Stars may always shine in a continually existing universe.

We may naively visualize the effect of a slowing progression of time by considering an object in motion. If time were to slow down and the second become longer, this object would move farther in a given time interval, which we would interpret as a higher velocity. Thus, slowing down the clock seemingly generates kinetic energy. Similarly, slowing the progression of time elevates temperature since the molecules in a gas or liquid move faster. In all instances, a slowing pace of time generates energy. This is illustrated in figure 14.

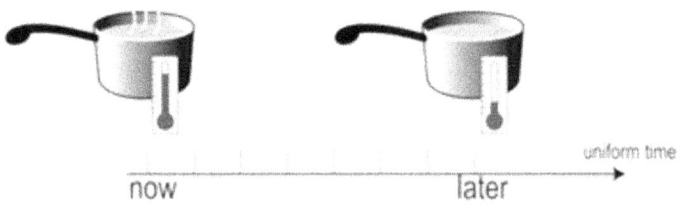

The temperature decreases if the pace of time is constant.

The temperature may remain the same if time slows down.

Figure 14: The pace of time influences temperature

However, this illustration does not take into account the expanding space, which dilutes the cosmological energy density. In the SEC, the energy density lost due to the expanding space exactly counteracts the energy generated by the expanding time, and the net cosmological energy is always zero. This balance eternally sustains the SEC universe as an open thermodynamic system that is continually in motion. It implies new physics that perhaps is not totally unexpected; since all epochs are equivalent in four-dimensional scale-expansion, it must imply conservation of the cosmological energy density.

Thus, the SEC is energized by the scale-expansion, and as we shall see below, its vacuum energy-momentum tensor (of GR) contains cancelling positive and negative components.

The universe is perpetually in a state of non-equilibrium, where energy generated by the slowing progression of time flows to the expanding space in a process that sustains all existence.

To further illustrate this new process we may think of keeping a slowly leaking balloon inflated. To maintain the inflation we may repeatedly add puffs of air. Similarly the "SEC balloon" is kept inflated by the incrementally increasing temporal metric that slows the pace of time while the expanding space causes cosmological energy to slowly leak away, for example via redshifted electromagnetic radiation.

Thermodynamics in the SEC

The SCM is facing a troublesome enigma with entropy. The closed universe of the SCM requires that the entropy should always increase, bringing with it steadily increasing cosmological disorder. But, rather than becoming more and more disordered, it appears that the universe of today is much more ordered than what it was just after the Big Bang.

On the other hand, the eternal SEC implies perpetual generation and dissipation of energy. Suns radiate with energy generated by nuclear processes, mainly the fusing of hydrogen to form helium. A sun will burn out and may eventually drift into the core of its home galaxy to be ejected as hydrogen gas, becoming material for new suns.

If the SEC model is correct, the energy of all matter could be induced via oscillating modulations of the spacetime metrics, which would imply that the cosmological scale-expansion is the origin of all energy in the universe. The radiating energy from suns and other sources is dissipated via the redshift. The cosmos is in thermal equilibrium, which explains the cosmic microwave background.

This means that the cosmos is a *thermodynamically open system* in which the net entropy may remain forever constant without increasing or decreasing.

Vacuum Energy in the SEC

With the SEC model, the cosmological energy-momentum tensor in vacuum evaluated in the cosmological reference frame does not disappear; the SEC theory implies that there is vacuum energy generated by the cosmological scale-expansion.

The cosmic energy tensor for the SEC theory is a diagonal matrix with four components:

$$CET = \begin{vmatrix} \dfrac{3c^2}{8\pi GT^2} & 0 & 0 & 0 \\ 0 & -\dfrac{c^2}{8\pi GT^2} & 0 & 0 \\ 0 & 0 & -\dfrac{c^2}{8\pi GT^2} & 0 \\ 0 & 0 & 0 & -\dfrac{c^2}{8\pi GT^2} \end{vmatrix}$$

Figure 15: The Cosmic Energy Tensor

In the figure G is the gravitational constant, T is the Hubble time, and c the speed of light

The temporal component T_{00} in the upper-left corner of the matrix corresponds to equivalent mass-energy density generated by the scale-expansion.

It equals Einstein's Critical Density of his 1917 paper on static cosmology.

It is the previously missing "Dark Energy", which could be nothing but spacetime energy induced by the temporal scale expansion.

The three spatial components model cosmological pressure. All three are negative, and each is equal to one third of the Critical Density. Therefore, the net cosmological energy density, which is the sum of all four components, disappears. The three spatial components play the same role as *Einstein's Cosmological Constant,* which he also postulated in his 1917 paper.

This explains the "accelerating cosmological expansion", which could be a misinterpretation based on the wrong SCM model. It is caused by the expanding spatial scale.

Therefore, both the Critical Density and the Cosmological Constant make their comeback in the SEC theory! And, like Einstein assumed in his static model, as perceived by an inhabitant the universe does not change with time.

Einstein found that his static cosmos model only could work if its mass density equalled the Critical Density and a Cosmological Constant also existed. On the other hand, in the SEC, the Critical Density and the Cosmological Constant do not have to be assumed or postulated to make the model work; they are both consequences of the scale-expansion.

The net vacuum energy in the SEC disappears, since there are canceling positive and negative components in the energy-momentum tensor. However, the scale-expansion generates spacetime curvature that accounts for the missing dark energy.

At the time of this writing, NASA and the Department of Energy (DOE), under the name NSPIRE, are calling for letters of application for membership in the Joint Dark Energy Mission Interim Science Working Group (JDEM), apparently in an attempt to resolve the enigmatic dark energy issue. I have tried to contact this program at two different e-mail addresses given at the NSPIRE Web site explaining that the dark energy problem simply disappears with the SEC model. However, NASA has not responded. This is another example where a large organization, driven by bureaucratic inertia, blindly continues down a given path. Perhaps people in control are motivated by other considerations than science. Of course, if there already is a solution for the dark energy problem, further research into its origin will not be needed.

Dark Energy

The mysterious Dark Energy is a hot subject in cosmology because it seems to comprise most of the energy throughout the cosmos. But how do we know that it really exists? It is needed in the SCM model to explain astronomical observations from huge distances. Presently the origin of this cosmological energy is unknown because the observable matter energy is merely a small fraction of the Dark Energy. The general consensus is that the cosmic energy density including Dark Energy ought to be the same as Einstein's Critical Density.

In the SEC model this Critical Density becomes a natural consequence of the cosmological scale expansion. Its origin is no longer unknown since it appears in the SEC model's energy-momentum tensor, making the cosmological scale expansion the primary energy source for the universe.

Einstein's Cosmological Constant

Einstein introduced the *Cosmological Constant* to solve a problem with his static universe of 1917. In such a universe, the gravitational attraction between galaxies would eventually cause all of them to converge to one central location. He solved this problem by assuming that in addition to gravitation, there is a repulsive force acting throughout the universe of just the right strength to counteract the gravitational contraction. This force appears as a constant in his GR equations—the Cosmological Constant. Later, when the expanding universe was discovered, he thought that his Cosmological Constant was no longer needed and regretted ever suggesting it.

But the Cosmological Constant is not dead—it has been revived from time to time for various reasons, most recently in the context of the supernovae 1a observations. In the SEC theory, it reappears together with the Critical Density. This combination generates positive and negative energy densities, which cancel each other out. As a result, there is no net energy in the SEC universe.

Vacuum Energy and the Cosmological Reference Frame

The question of vacuum energy and gravitational field energy are closely related. People familiar with GR take great pains to explain that the spacetime vacuum field energy must disappear. In the SCM, this must be true because if there were any vacuum energy density, it would be different as seen from different inertial frames. This would be impossible if inertial frames were physically equivalent. In the SCM, the only solution to this dilemma is that the vacuum density must disappear in all coordinate frames.

But in the SEC, where there is a cosmological reference frame, this is no longer the case. Measured in this reference frame, the vacuum energy-momentum tensor is the cosmic energy tensor, and the net energy disappears. However, this is no longer true for coordinate frames in motion relative to the reference frame. This explains why there is cosmic drag in the SEC universe. As we shall see in the next section, the SEC theory also implies that the energy of a gravitational field is negative. Moreover, this negative energy exactly balances the positive energy of gravitating matter.

Again, there is no net energy in the SEC universe!

The Zero-Point Energy Enigma

The *zero-point energy (ZPE) of the cosmological vacuum* is yet another unresolved problem with the SCM. ZPE is the lowest possible energy that a quantum mechanical physical system may have; it is the energy of its ground state. All quantum mechanical systems undergo fluctuations even in their ground state at zero absolute temperature and have associated zero-point energy, a consequence of the Heisenberg uncertainty principle.

Vacuum energy is the combined zero-point energy of all the fields in space, which includes the electromagnetic field. The problem with the SCM is that the predicted ZPE is enormously large when adding the contributions from all possible oscillatory frequencies. Based on quantum mechanics it is estimated that the ZPE density is a factor 10^{110} larger than the energy density at the center of the Sun! In the SCM, this is unexplainable. Clearly this shows that we do not really understand the physics of the ZPE.

On the other hand, if the vacuum oscillations are in the *scale* of spacetime rather than in coordinate space, the net contribution from each mode of oscillation will disappear because the corresponding energy-momentum tensor takes the same form as the cosmic energy tensor; its diagonal components sum to zero and the net contribution for each oscillatory mode disappears.

In the SEC the zero-point vacuum energy disappears!

We have failed to realize that the ground state of quantum theory may refer to oscillations in the metrical scale rather than to quantum field waves in space. As we shall see in chapter 8 and appendix IV, this also agrees with the finding that the domain of the quantum mechanical wavefunctions is the metrical scale of spacetime rather than the four coordinate dimensions of space and time.

Gravitation in the SEC

The question of gravitational field energy has been investigated by many including Einstein, who was puzzled by the fact that the gravitational field energy disappears in the SCM. He still felt that there must be some way of expressing gravitational field energy and invented a "pseudo-tensor" for this purpose. However, this tensor does not really belong in GR since it does not transform like an ordinary tensor. Other investigators have since proposed different gravitational pseudo-tensors, but recent investigation has shown that all these pseudo-tensors can be made to disappear with special choices of coordinates [Neto and Trajtenberg, 2000]. Therefore, these pseudo-tensors are nothing but smoke and mirrors; they do not have any physical meaning.

Although there can be no vacuum energy or gravitational field energy in the SCM, we know that the cosmological vacuum contains dark energy and therefore that our understanding lacks something very important. The SEC theory resolves these issues in a simple and straightforward manner. The problem can be traced all the way back to Galileo's assumption that relative velocities for freely moving objects will not diminish over time. This became Newton's first law of motion, which in the context of special relativity must imply that inertial frames are equivalent and that there is no cosmological reference frame. But in the SEC, there is cosmic drag, which violates Newton's first law and defines a cosmological reference frame.

The material summarized in this section may be found in [Masreliez, 2004c].

Schwarzschild's Solution

In standard physics, there is an exact solution to Einstein's GR equations, *provided that all components of the energy-momentum tensor disappear*. This solution was discovered by Karl Schwarzschild and published in 1916, the year of his premature death at the age of 42. His solution is remarkably simple and is the basis for the belief that black holes exist, which unfortunately over the years has been accepted as a proven fact. But, black holes are purely hypothetical objects, based on the assumption that the energy-momentum tensor of GR disappears and that there is no cosmological scale-expansion. If this is not the case, like in the SEC theory where there is vacuum energy induced by the scale-expansion, black holes do not exist, even theoretically.

Schwarzschild's solution, as well as Newton's law of gravitation, implies that in the far field far away from gravitating matter, the gravitational potential takes the following well-known form as a function of radial distance:

$$P(r) = \frac{G \cdot m}{r}$$

As usual, G is the gravitational constant and m the gravitating mass.

Since the days of Newton, this relation has been puzzling since it implies that the combined gravitational potential from all the matter in a homogenous infinite universe must be infinitely large and that the gravitational force acting on any particle (pulling it in all directions) is also infinitely large. Of course, this is quite disconcerting. Over the years, many have attempted to resolve this puzzle. Perhaps there is a limited amount of matter in the universe? Perhaps the observable cosmos is only an island in empty space?

Some people have proposed a modified potential that diminishes its action at large distances. The most well-known is probably the Yukawa potential with an exponential "roll-off" factor:

$$P(r) = \frac{G \cdot m}{r} e^{-r/R}$$

This attenuates the potential at some large distance R. However, this is a hypothetical ad hoc solution.

There is another already mentioned strange property of Schwarzschild's solution: it assumes that the energy-momentum tensor in vacuum disappears everywhere, even very close to gravitating matter. This means that the cosmological vacuum does not contain gravitational energy, which conflicts with the conclusion based on other considerations that the gravitational field energy ought to be negative.

The SEC Solution

If black holes do not exist, we may wonder what happens during gravitational "collapse." *However, total gravitational collapse is prevented by the cosmological scale-expansion.*

In order to understand how this might be possible, we will assume that the line-element in GR that models a spherically symmetric field includes two modifications compared to the corresponding SCM line-element:

- There is cosmological scale-expansion acting on all four metrics of spacetime.

- The energy-momentum tensor for cosmological vacuum no longer disappears but equals the cosmic energy tensor of the SEC model.

When doing this, we find that an exact solution to the GR equations no longer exists! There is an approximate solution similar to the Schwarzschild solution, but it does not hold for very small and very large distances. How should we interpret this?

I think it tells us that in order for an exact solution to exist in the SEC, the vacuum energy-momentum tensor must change in the vicinity of matter.

Thus, the presence of matter changes the vacuum energy density.

Moreover, we can guess how the energy-momentum must change since we know that the solution should be almost exactly the same as the Schwarzschild solution at intermediate distances. At large distances we then find that the gravitational potential effectively disappears close to the Hubble distance; regions even farther away do not influence us gravitationally. Furthermore, assuming constant cosmological mass density, the gravitational potential from all matter in the universe is finite, and the total gravitational field energy is negative; *it equals* $-mc^2$.

Not only does the SEC model limit the range of gravitation, but the gravitational field energy also equals that of gravitating matter, but with the opposite sign. Thus the presence of matter does not contribute to the net cosmological energy.

The SEC Near-Field Solution

In the SEC, black hole formation is prevented by the scale-expansion. Since the vacuum energy-momentum tensor does not disappear, Schwarzschild's exterior solution, which implies the possibility of black holes, no longer exists. We can show this by applying GR in a rather technical discussion, which is beyond the scope of this book. The interested reader might find the details of this development in my paper [Masreliez, 2004c]. However, I will summarize the main conclusions here.

In the SCM, the temporal metric of Schwarzschild's solution becomes zero and the radial metric becomes infinite at a radial distance called the event horizon. This is the distance that signifies the radius of a black hole. Since the temporal metric goes to zero, it means that the progression of time stops at the event horizon. These are well-known aspects of a black hole. When approaching the event horizon, the solution for the SEC line-element closely follows the same trend, but with the difference that gravitational field energy density becomes sharply negative close to the event horizon. This suggests that inward motion is prevented since negative energy will cause gravitational repulsion rather than attraction.

It is possible to find an approximation to the SEC solution that holds very close to the event horizon and use this solution to investigate the trajectory of a particle falling inward. We find that it never reaches the event horizon and therefore that a black hole cannot form. Technically, *the event horizon is a singularity in the SEC;* it is a forbidden radial distance at which matter cannot exist, while in the SCM an object may fall straight through the event horizon.

The SEC Resolves Gravitational Puzzles

It is intriguing that the tiny vacuum energy density and the very slow scale-expansion of the SEC theory both limit gravitational action at cosmological distances and prevent black hole formation at very small distances. Two main enigmas with the current theory of gravitation "magically" disappear. Also, the presence of a cosmological reference frame allows the existence of a negative gravitational field energy that balances the gravitational matter energy.

Epistemological Implications

I have presented these philosophical and conceptual implications of the SEC theory with some trepidation, realizing that they may be hard to digest, in particular for the reader well versed in physics. But I have found it impossible to present the theory in a piecemeal way without leaving too many unanswered questions.

It is tempting to ignore the SEC model altogether rather than having to face the possibility that SCM, our current paradigm, falls short, but unfortunately the SEC model cannot be accommodated merely by adjusting the SCM; it implies a major revision of our worldview.

CHAPTER 7

Motion and the Origin of Inertia

The origin of the inertial force has remained a mystery ever since the dawn of Western science. Inertia is the tendency to resist acceleration by a force given by Newton's famous second law of motion, $F = am$, which was postulated but not derived by Newton. Newton assumed that the inertial mass appearing in this expression is the same as the gravitational mass appearing in his law of gravitation, an assumption that has been verified by numerous experiments, and which formed the basis for Einstein's GR theory. Therefore, inertia and gravitation appear to be closely related and might have the same origin. And since gravitation, according to GR, is a curved spacetime phenomenon, inertia should also be associated with spacetime curvature; *but this connection has, in the past, been missing.*

Nobody has understood why a particle in empty space should resist acceleration. This chapter shows how inertia like gravitation could be caused by spacetime curvature, an explanation that unfortunately will require conceptual reassessment of special relativity.

The Mystery of Motion

General relativity is one of the flagships of physics. Einstein thought of it as primarily a theory of gravitation. Its main advantage over past theories was that it provided an explanation of the gravitational force. According to GR theory, gravitation is caused by spacetime metrics that change with spatial location (spacetime curvature). In particular, the gradient in the temporal metric causes the gravitational force here on the surface of the Earth.

However, GR is based on static 4D geometry that does not model the *dynamic process* of motion. Conceptually, it may be viewed as describing a fixed geometry much like we describe a 3D object, like a cube or a sphere, in 3D space. In GR, there is *no process of motion* in the sense that it does not describe how an object moves from one location to the next.

As already mentioned, motion is in GR modeled by a "world-line," which is a one-dimensional trajectory through 4D spacetime. Each point on this trajectory corresponds to a certain location in space and time. We might say that GR records the history of motion for an object that allows us to know where it was at different times in the past. However, this history does not end with the present time according to GR; it also continues unabated into the future without distinguishing between the past and the present. If you think this is strange you are right. Although GR might be correct in certain aspects, it clearly cannot be the last word in describing the world.

This was not a problem as long as GR was used to model static gravitational fields. In agreement with this property of GR Einstein in 1917 suggested a static cosmological model dominated by gravitation from matter. This model describes a universe with fixed GR geometry.

However, unfortunately GR was also later used to model motion in the form of the cosmological expansion, which was a bad mistake. Someone should have realized that motion is *a process*, which cannot be modeled by (static) geometry. The main reason for this mistake is that the process of motion is not yet fully understood. In fact, motion remains a mystery and has always been mysterious to people who have tried to analyze it in detail.

It is not difficult to see why. Motion means that an object is at one location at one point in time and at another location some time later, but how can this happen if the object is rigid? It seems that it must stretch and perhaps move forward like an inchworm, because it seems impossible that an object with fixed dimensions can move at all. This means that either no objects exist with absolutely fixed dimensions, or perhaps that all motion occurs in tiny incremental steps.

This also puzzled the ancient Greeks; Zeno's paradoxes about the nature of motion are well known. The most famous of Zeno's arguments is the one about Achilles and the Tortoise. The original goes something like the below, which can be found at www.mathacademy.com/pr/prime/articles/zeno_tort/index.asp:

"The Tortoise challenged Achilles to a race, claiming that he would win as long as Achilles gave him a small head start. Achilles laughed at this, for of course he was a mighty warrior and swift of foot, whereas the Tortoise was heavy and slow.

"How big a head start do you need?" he asked the Tortoise with a smile.

"Ten meters," the latter replied.

Achilles laughed louder than ever. "You will surely lose, my friend, in that case," he told the Tortoise, "but let us race, if you wish it."

"On the contrary," said the Tortoise, "I will win, and I can prove it to you by a simple argument."

"Go on then," Achilles replied, with less confidence than he felt before. He knew he was the superior athlete, but he also knew the Tortoise had the sharper wits, and he had lost many a bewildering argument with him before this.

"Suppose," began the Tortoise, "that you give me a 10-meter head start. Would you say that you could cover the 10 meters between us very quickly?"

"Very quickly," Achilles affirmed.

"And in that time, how far should I have gone, do you think?"

"Perhaps a meter—no more," said Achilles after a moment's thought.

"Very well," replied the Tortoise, "so now there is a meter between us. And you would catch up that distance very quickly?"

"Very quickly indeed!"

"And yet, in that time I shall have gone a little way farther, so that now you must catch that distance up, yes?"

"Ye-es," said Achilles slowly.

"And while you are doing so, I shall have gone a little way farther, so that you must then catch up the new distance," the Tortoise continued smoothly.

Achilles said nothing.

"And so you see, in each moment you must be catching up the distance between us, and yet I—at the same time— will be adding a new distance, however small, for you to catch up again."

"Indeed, it must be so," said Achilles wearily.

"And so you can never catch up," the Tortoise concluded sympathetically.

"You are right, as always," said Achilles sadly—and conceded the race."

We must face the fact that motion still is as mysterious as it was two and a half thousand years ago, although many scientists will refute this claim. To those who do refute it, I have only one question: "What causes the progression of time?" If they cannot answer this simple question, they do not really understand motion.

Here, I approach this puzzle from a new angle, which makes use of the SEC model with its DIST process. I show that it offers a new perspective on motion that could lead to a deeper understanding with potentially important consequences, see further Appendix VI.

A Constant Velocity of Light?

One of the main unresolved questions facing physics at the beginning of the twentieth century concerned the nature of light. A famous experiment by Albert Michelson and Edward Morley showed that the velocity of light seems to be the same regardless of the motion of the Earth relative to the aether that everyone assumed existed [Michelson and Morley, 1887]. In his SR theory Einstein bypassed this question and simply assumed that the speed of light is constant but he never tried to explain what physically happens to a particle when accelerating from one velocity to another. Implicit in his thinking was the idea that material bodies are well-defined fixed entities that may be used as references regardless of motion. Einstein often referred to rigid bodies and ideal clocks. This has caused a lot of confusion.

However, if material particles were to consist of resonating waves in the metrics of spacetime, they would not exist independently of spacetime but would somehow be integrated with it. Therefore, using material bodies as independent references is questionable. If the existence of particles results from various modes of metrical oscillation, it is possible that they may adjust their metrical properties depending on their velocities in order to locally keep the speed of light the same in all directions. The reason for this metrical "morphing" could be that it will preserve conditions needed for a particle's existence. Thus, instead of adjusting its dimensions, the moving object might instead adjust its local spacetime geometry, preserving the conditions necessary for its existence! As a consequence the speed of light could locally remain the same for all moving particles.

Since an observer always uses the spacetime geometry in her local frame, which may adjust to her motion, the relative speed of light could always remain the same locally as measured with these local metrics. However, an observer in another inertial frame, who uses different metrics, may *not* find that the light is constant and isotropic in a different inertial frame. Einstein actually used this in his derivation of special relativity.

This important shift in perspective is in agreement with the essence of relativity; locally the world looks the same simply because the geometry of space and time adjust to accomplish this. We may say that the velocity of light is not constant, but rather that *the morphing spacetime geometry causes it to become constant*. With this new possibility comes the realization that *the spacetime geometry may change for a particle that accelerates, and that this change might be what causes the inertial force.*

By the classical way of visualizing motion, we consider a rigid coordinate frame that moves in relation to another rigid frame with coordinates that are related via the classical Galilean transformation. However, with the discovery that the speed of light remains constant and the introduction of the Lorentz transformation came the notion that particles somehow change in a relative sense when in motion.

Here, I propose that particles always remain the same in relation to their local reference frame, but that a frame's geometry changes during motion relative to other inertial frames. In other words, acceleration might curve spacetime while locally keeping the velocity of light the same.

The DIST Process and Inertia

We saw that the dynamic incremental scale transition process, whereby the metrics of spacetime change semi-continuously, may describe the cosmological scale-expansion and that it will result in a superior cosmos model. The main difference between the DIST process and other previously considered dynamic processes is that it will preserve all properties of the 4D spacetime relative to observers who participate in the process. Therefore, it is a process in scale "beyond space and time," which locally does not change the 4D geometry.

The cosmological expansion may be seen as acceleration in time; the scale increases by the same small fraction each second and will thus accelerate geometrically (exponentially), which will have observable effects—for example, the cosmological redshift, cosmic drag, and dark energy. Yet the world we live in always remains the same.

According to SR and GR, coordinate frames moving at constant velocities (inertial frames) are, in the absence of gravitational fields, geometrically "flat" as represented by the Minkowskian line-element. This means that an observer in an inertial frame will find that all laws of physics hold true in the local frame. Inertial frames are physically equivalent in all respects; in particular, the speed of light is the same in them.

Although SR does not model acceleration, it seems strange that all inertial frames should have exactly the same Minkowskian line-element, because if inertia were a phenomenon akin to gravitation it should, like gravitation, be caused by spacetime curvature. However, this is not possible if the geometry of all inertial frames is one and the same; then spacetime curvature cannot explain the inertial force. Therefore, it appears that we may have missed something important.

The fact that the DIST process replicates the same spacetime geometry at different scales suggests that acceleration in space might also change the 4D scale of the line-element and that therefore the DIST process also might apply to spatial acceleration.

I investigated this possibility by applying an arbitrary dynamic scale-factor to the Minkowskian line-element, which I assumed changes with location in space just like the spacetime metrics do in a gravitational field [Masreliez 2006a, 2007a, 2008]. I wondered if an accelerating particle might experience spacetime curvature like a particle does in a gravitational field, which possibly could explain inertia as a gravitation-type phenomenon.

When using this assumption in Einstein's geodesic equation of GR, I found, to my delight, that a certain scale factor exists for which all accelerating motions, regardless of magnitude, become spacetime geodesics.

With this scale factor, the geodesic equation of GR becomes an identity.

Regardless of the magnitude and direction of its acceleration, an accelerating particle will always move on a geodesic of GR! I also found that this scale factor does not depend on location, as I initially had assumed, but only on the velocity. This was a good sign because it implied that the scale is constant if the velocity is constant, with the further implication that all laws of physics would be conserved in inertial frames due to scale equivalence! Consequently, inertial frames would be physically equivalent just like they are in SR and the speed of light would be constant and the same in them.

This result strongly suggests the following:

Acceleration curves spacetime and induces the inertial force. Spacetimes of different velocities differ in relative scale.

However, although inertial frames are physically equivalent this would disagree with SR's Lorentz transformation, which does not change the scale.

The scale-factor that would explain inertia is given by $S(v) = 1 - (v/c)^2$, where v is the relative velocity and c is the speed of light. I call this the "inertial scale-factor" and the corresponding Minkowskian line-element scaled by this dynamic scale-factor the "inertial line-element."

People familiar with SR will recognize this inertial scale-factor, since the square root of $S(v)$ appears prominently in SR in the contexts of time dilation and length contraction. In fact, we find that there is a close connection between the new theory of inertia and SR.

How Motion on a Geodesic Explains Inertia

A freely falling particle in a gravitational field follows a trajectory given by the geodesic of GR. No force acts on such a freely falling particle. Therefore, as perceived by a freely moving (or falling) observer, free motions in empty space or in a gravitational field are very similar. This important insight became a major inspiration for Einstein during his development of GR.

However, an object here on the surface of the Earth is prevented from falling by an opposing and supporting force from the ground, which points in a direction opposite to that given by the geodesic. This supporting force, which equals the weight of the object, counteracts the gravitational force which has a direction given by the geodesic.

Now let us compare this to the situation where an object is accelerated by applying an external force. This applied force may be compared to the supporting force from the ground; it prevents the object from "freely falling", i.e. moving at constant velocity. Like with the gravitational field, where the supporting force is balanced by the gravitational force, there is an inertial force counteracting the applied accelerating force. We explain the gravitational force as being due to spacetime curvature. Likewise, we may explain the inertial force as being due to spacetime curvature if *the acceleration takes place on a geodesic.*

With the inertial line-element, this relationship holds true for an arbitrary accelerating force regardless of its magnitude and direction, since the geodesic equation is an identity. Therefore, the inertial force induced by the inertial field always balances the applied force. In other words, the magnitude of the inertial field is always such that it induces just the right inertial force.

This is illustrated in figure 16. Note that on the right-hand side the acceleration vector is pointing upward, indicating that an object at rest on the surface of the Earth actually is accelerating in relation to its natural motion on its geodesic, which is free fall. The reaction force from the ground, which prevents it from falling, plays the same role as the applied accelerating force shown on the left-hand side. In both cases, spacetime resists acceleration with a balancing inertial or gravitational force.

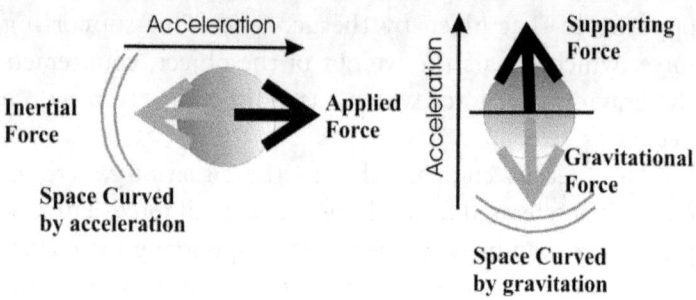

Figure 16: Inertia and Gravitation curves spacetime

A new and important aspect should be noted here. With inertia we are dealing with a gravitational field induced by a *dynamic* scale factor, which changes with location and with the applied force. It automatically balances the applied force regardless of its magnitude or direction. This is different from gravitation, which is typically modeled by a static force field.

Other Attempts to Explain Inertia

Over the years, since the time of Newton, many attempts have been made to explain the origin of inertia. Often, inertia is believed to be intimately connected to the existence of a cosmological reference frame, since without such a reference frame it is hard to understand how acceleration may arise at all. In other words, if there is no spatial reference frame, how can a particle even sense that it accelerates?

We saw that Ernst Mach suggested that distant matter in the cosmos defines a reference frame (Mach's principle) and that accelerating particles somehow sense acceleration relative to this cosmological reference frame. However, this cannot explain why a particle instantly resists acceleration via the inertial force. It seems that local space must be influenced by distant matter, but nobody knows how this comes about.

More recent approaches attempt to explain inertia as arising from interaction with the zero-point field (ZPF). According to QT the ZPF represents vacuum energy that permeates all space. Due to Heisenberg's uncertainty principle, the lowest possible energy state in a vacuum is positive rather than zero. The cosmological vacuum is filled with field energy oscillating at many frequencies, and it seems possible that this zero-point field somehow interacts with acceleration and might explain inertia.

The idea that particles are different from space and time is common to these explanations. A particle is envisioned as moving in some kind of vacuum medium or field (aether) similar to moving in air or water, and one speculates that inertia somehow arises from interaction between a passive particle and this medium or field.

The new explanation to inertia advocated here resolves the question of a spatial reference by assuming that a particle actively participates in its motion by adjusting its local scale. At each instant the reference frame is provided by the particle itself; it compares its immediately preceding state to its current state in a stepwise manner via the DIST process. The origin of inertia is therefore a local phenomenon for each particle. This agrees with our perception of acceleration as being caused by a changing velocity relative to what it was at the immediately preceding instant. At each instant, the particle and its instantaneous geometry define the reference for future action. Although this to some extent may be modeled by SR via repeated velocity boosts, SR does not allow inertia to be explained as a curved spacetime phenomenon.

The reason is this: consider a sequence of very short velocity boosts separated by equally short intervals of constant velocities. This scenario is often used to model acceleration in SR. However, according to SR, each segment of constant velocity has the same Minkowskian geometry, which means that because the geometry does not change inertia cannot be modeled in GR as a curved spacetime phenomenon similar to gravitation. Yet we know that the inertial force and gravitational force are of the same kind and therefore that they should have the same origin.

The DIST process resolves this conundrum by allowing the scale of spacetime to change during acceleration while preserving the same Minkowskian geometry within the segments with constant velocities. An accelerating particle senses the changing scale of spacetime by the continuous portion of the DIST process, and retains its Minkowskian geometry by discretely adjusting its scale.

The DIST process allows dynamic action via the metrics of spacetime in a process not recognized by standard physics. This explains why the origin of inertia previously has remained elusive.

A Few Conceptual Problems with Special Relativity

With his paper on SR in 1905, Einstein abolished the aether, which until then had defined an absolute cosmological reference frame. According to SR, all coordinate frames moving with constant relative velocities (inertial frames) are considered physically equivalent and the speed of light is constant and equal in all of them, a concept that is hard to understand if photons are particles that move at the speed of light. SR also predicts that an observer in an inertial frame will see clocks in other frames run slower. The infamous Twin Paradox is a consequence, where twins who part and travel in relative motion each conclude that the other twin ages slower and should be younger when they later reconvene.

The Twin Paradox has been heatedly discussed at great length since the introduction of SR but has, in my opinion, never been satisfactorily resolved. Today, this discussion is still very much alive on the Internet.

Einstein based this SR theory on the following two postulates:

1. The laws of physics hold true in all inertial frames.

2. The speed of light is constant, isotropic, and the same in all inertial frames.

It is commonly believed that these two postulates *imply* the Lorentz Transformation and that they necessarily lead to SR. However, as we shall see, this is not true.

Einstein implicitly also assumed that coordinate increments of the transformed, moving coordinates have the same meaning as those of the stationary reference frame so that coordinate increments may be directly compared.

However, in order to achieve equivalence it is not necessary that the transformed line-element is Minkowskian, because inertial frames might have different scales and be scale-equivalent. It is interesting to note that there is another transformation similar to the Lorentz transformation, which changes the scale of spacetime and is consistent with the inertial line-element.

Woldemar Voigt's Transform

In 1887, the German physicist Woldemar Voigt published a paper proposing a coordinate transformation between inertial frames *that corresponds to the inertial scale-factor* when the velocity is constant.

In one of my papers I independently re-derived this transformation, calling it the Scaled Lorentz Transformation, being unaware of the Voigt Transformation (VT) [Masreliez, 2007a]. The VT becomes identical to the Lorentz Transformation if the factor γ is applied to all four of its relations:

$$\gamma = \frac{1}{\sqrt{1-(v/c)^2}}.$$

The inverse of the VT differs from the forward transformation. The VT is therefore asymmetrical in that the forward transformation differs from its inverse. However, for the LT, the forward and inverse transformations coincide.

Thus, a transformation that formally implements the inertial line-element for constant velocities actually preceded the Lorentz transformation!

Apparently Hendrik Lorentz did not know about this aspect of Voigt's work; he is on record as saying that if only he had known about it, he might have taken Voigt's transformation into his theory of electrodynamics rather than develop his own, as evidenced by the following letter appearing in [Ernst and Hsu, 2001, (p. 214)]:

H. A. Lorentz to W. Voigt   *Leiden, July 30, 1908*

Dear friend,

I would like to thank you very much for sending me your paper on Doppler's principle together with your enclosed remarks. I really regret that your paper has escaped my notice.

I can only explain it by the fact that many lectures kept me back from reading everything, while I was already {sic} glad to be able to work a little bit.

Of course I will not miss the first opportunity to mention that the concerned transformation and the introduction of a local time has been your idea.

Sincerely,
Your H. A. Lorentz

The VT differs from the LT by a constant scale-factor, which in the LT has the effect of restoring the Minkowskian line-element, thereby hiding the scale adjustment that models inertia. The two transformations are physically equivalent and work equally well in Maxwell's equations as well as in physics in general, since they are scale-equivalent. It is ironic to note that if Lorentz had adopted the VT, Einstein might perhaps have chosen the VT instead of the LT in his SR paper, and the origin of inertia might have been found a long time ago!

A Previously Overlooked Conceptual Difficulty

The use of coordinate transformation like in the VT or LT implies an ambiguity that was briefly alluded to in the introduction above. In modeling motion it is unclear how coordinates of a moving frame relate to those of a stationary frame.

Assigning coordinates is a convenient way of specifying locations in space and time; we may think of the world as spanning three spatial coordinates and one temporal coordinate. We are free to assign these coordinates in various ways; in fact, it can be done in infinitely many ways. And if we interpret the coordinates as giving locations in spacetime, we may assign physical meaning to coordinate increments. Together, they specify the *geometry* of spacetime, as is done by the line-element of GR.

The coordinate transformations mentioned above attempt to model motion by considering moving points in the stationary frame. *However, these moving points do not belong to the stationary frame's geometry.*

In personal communication a mathematical logician has forcefully argued that points do not move in geometry.

In geometry, no point moves as it breaks the continuum of geometric space. Geometric transformations are not physical motions. Relationships between points and the concept of motion are different things.

In other words, in mathematics geometry is fixed; points do not move. Therefore, if we want to use geometry and assign meaning (metrics) to coordinates, we are constrained to relations between fixed coordinate locations in a particular frame. Each frame has its own locally fixed geometry, and the locations of points in this geometry are fixed. The geometries of two different frames in relative motion are conceptually unrelated; their relative motion cannot be described by geometry.

This puts into question the use of coordinate transformation in modeling motion.

In his SR paper, Einstein (1905) attempted to bridge this conceptual gap. Using light signals he assigned the coordinates in the moving frame so that his two postulates were fulfilled. However, this procedure left him with an undetermined constant—*the scale*. He determined the value of this constant by assuming that each inertial frame has identical local geometries and that *as perceived by a stationary observer, the geometry of a frame in relative motion belongs to the same spacetime geometry as that of the stationary reference frame.* But, as we have seen, this assumption cannot be justified since moving points do not belong to the stationary reference frame's geometry. Therefore, it is not impossible that the scale of a moving frame will appear to be contracted.

This is a previously unrecognized conceptual difficulty with SR, which is causing puzzling problems—for example, the Twin Paradox.

In this context it is enlightening to recall Aristotle and his version of Zeno's Arrow Paradox:

1. *When the arrow is in a place just its own size, it's at rest.*

2. *At every moment of its flight, the arrow is in a place just its own size.*

3. *Therefore, at every moment of its flight, the arrow is at rest.*

This ancient insight is remarkable since it recognizes that a moving object always remains fixed in its local geometry.

Two Different Approaches to Modeling Motion

In classical as well as in modern physics, motion is usually modeled in terms of coordinate locations that change with time, and transformations such as the LT and the VT relate these locations expressed in different coordinate frames. Thus, both these transformations relate positions in space and time so that positions in the reference frame correspond to positions in the moving frame and vice versa in one-to-one correspondence.

Furthermore, SR is based on the LT and assumes that these two frames have identical coordinate metrics, so that coordinate increments in one frame may be directly compared to those in the other frame.

This is a crucial but often overlooked implicit assumption that doesn't necessarily hold true.

Thus, the assumption is made that the meaning of the coordinates *observed in the moving* frame is the same as for those of the stationary frame. This overlooks the possibility that coordinate increments in the moving frame obtained by coordinate transformation might not have the same meaning as those in the stationary frame. Coordinate transformation might offer different "perspectives" that depend on the observer's frame of reference. If a stationary observer's perspective of a moving frame were to differ from the perspective of a local observer in the moving frame, it could explain the Twin Paradox. In other words, what we see in a moving frame might not be what actually goes on in it. Each inertial frame may be physically the same, which would agree with Einstein's first postulate: the laws of physics hold true in all inertial frames, yet moving frames might be experienced differently.

The VT and LT may also both be viewed as continuous coordinate transformations in GR, relating geometries rather than modeling motion. This geometric point of view allows us to compare coordinate increments, and the VT implies that intervals in space and time appear to be contracted in a moving frame. In particular, time seems to progress slower, which models time-dilation.

However, had we instead interpreted the VT as a generalization of the classical Galilean transformation, time dilation would not be evident because with this interpretation the VT does not allow us to relate time intervals *at a fixed location* in the reference frame to time intervals *at a fixed location* in the moving frame.

On the other hand, if the LT instead is seen as a coordinate transformation in GR, allowing us to compare coordinate increments in the two frames, the moving frame will have the same Minkowskian line-element as the stationary frame and their clock rates will be the same, which means that time dilation disappears.

Therefore, the geometric interpretation favors the VT.

Woldemar Voigt's objective with his transformation was to preserve wave equations in the aether, while Lorentz desired to preserve Maxwell's equations, and Einstein based SR on his two postulates (which also hold for the VT). However, from the geometric point of view, the reason for the success of the VT and LT could be that they both preserve the character of spacetime, because they are scale-equivalent. Therefore, the primary role of both the LT and the VT could be to implement a scale-equivalent Minkowskian coordinate transformation in GR. However, only Voigt's transformation is consistent with a dynamic metric that models Inertia.

It is not surprising that the LT and the VT were seen as modifications of the classical Galilean coordinate transformation, because at the time GR was not yet available. However, after GR had been introduced in 1915, it would have been possible to reassess SR; in particular, we could have questioned the use of coordinate transformation in modeling motion. However, by this time SR was well established and formed the basis for the new physics of relativity. After Einstein's fame further increased with the introduction of GR, the position of SR was solidified. However, ever since its introduction, dissenting voices against SR have risen and continue even today; the debate is still very much alive on the Internet.

As we shall see, this criticism is justified.

Resolving the problem with SR and the Twin Paradox

With this background we are ready to address the Clock Paradox, popularized as the Twin Paradox by Paul Langevin [Langevin, 1911].

This strange implication of SR has been discussed at great length ever since its introduction and several proposals on how to resolve it have been put forward over the years. In my opinion, the Twin Paradox is a fundamental conceptual inconsistency of SR that has been swept under the rug, so to say; people have simply learned to live with it. I think most people in science recognize that since both twins cannot be right in that the other twin returns younger, the view from a local stationary frame must differ from the views from frames in motion. If this is true there is conceptual problem with SR since it gives no physical explanation for this relativistic difference in perception. Why does the observed time seem to run slower in a moving frame if the clock in this frame does not run slower?

A few uncompromising individuals—for example, the brave and persistent Herbert Dingle, repeatedly challenged SR by claiming that something is seriously wrong with the theory. This quote is taken from *Wikipedia's* entry on Herbert Dingle:

> *"Dingle became a professor of Natural Philosophy at Imperial College in 1938, and was a professor of History and Philosophy of Science at University College London from 1946 until his retirement in 1955. Thereafter he held the customary title of Professor Emeritus from that institution. He was one of the founders of the British Society for the History of Science, and served as President from 1955 to 1957. He founded what later became the British Society for the Philosophy of Science as well as its journal, the* British Journal for The Philosophy of Science."

Dingle was obviously well qualified and knew SR quite well since he wrote the essay *"Relativity for All"* (1922) and the monograph *"The Special Theory of Relativity"* (1940).

However, he later became convinced that SR wasn't the final word and tried to make the science establishment aware of this troublesome situation. But, unfortunately, SR had then been elevated from a mere theory of physics to become the gospel of absolute truth that should not be challenged. Dingle's repeated efforts were rejected, and his warnings ignored. In frustration Dingle then published the book *"Science at the Crossroads"* (1972), in which he strongly expressed his opinions and misgivings. He claimed that it would be a serious mistake to ignore this problem with SR, since much of modern science depends on this theory.

As we shall see, Dingle was right.

Nowadays, many believe that the Twin Paradox has been resolved one way or the other, although it appears that a universally accepted resolution still is missing; at least I have not seen one.

Let's take a closer look at the Twin Paradox:

In SR, problems arise when comparing observations made by observers in different inertial frames. Twin Paradox, popularizes this problem. Strong experimental evidence exists that the pace of time in a frame in motion actually appears to slow down. As experienced by a stationary observer, time in a moving frame appears to run at a slower pace compared to time observed in a stationary reference frame. Based on this confirmed fact, it is tempting to conclude that the twin who accelerates away and later returns ages slower and therefore returns younger. However, the consensus is that acceleration does not influence clock rates.

SR postulates clocks in inertial frames must progress at the same pace. Therefore, each of the twins should experience the same local pace of time regardless of their relative motion, and the two clocks should agree when reconvening, which contradicts the fact that the pace of time appears to go slower in moving frames. Some argue that there is a difference between the twins because only one of them accelerates. However, if both twins were to accelerate symmetrically in opposite directions and then coast for some time before symmetrically reconvening, they would during their relative motion each see the other's clock be running slower, and thus, the Twin Paradox would still remain un-resolved. Since their motions are symmetrically the same their clocks must run at the same pace; there is obviously an inconsistency, which is the essence of the paradox.

Let's now see how this mystery may be resolved.

We saw that the use of coordinate transformation is problematic when describing motion because moving points do not belong to a stationary frame's geometry. As a consequence the meaning of the transformed coordinates is questionable; it might not be the same as in the stationary frame and therefore we cannot compare coordinate inter-vals. However, the use a GR transformation allows us to compare coordinate intervals. Therefore, if we use a coor-dinate transformation to relate the geometries in moving frames it should be a transformation in GR. In this case the coordinates of moving frames are related by the previous-ly defined geometric approach. However, this means that we should use Voigt's transformation rather than Lorentz's because only the VT captures time-dilation and length-contraction. This would mean that the observed scale of spacetime is contracted in a relative sense for frames in mo-tion, but that this does not influence the local conditions in each frame.

This interpretation has three major advantages:

1. It resolves the Twin Paradox

2. It allows a common temporal reference

3. It explains the inertial force

It resolves the Twin Paradox since the diminished scale in a moving frame allows an *observed* clock to run slower without influencing its local pace in the moving frame. It allows a common temporal reference because moving frames are related via Voigt's transformation instead of Lorentz's. And, it explains inertia as being a curved spacetime phenomenon induced by the dynamic spacetime scale.

However, it has one "drawback" if you believe that the world may be described by standard physics because it implies that the relative perspective between moving frames involves an additional aspect in the form of a dynamic spacetime scale. Hence, it is no longer sufficient to model relative motion merely via changing locations in four-dimensional spacetime; the additional scale-dimension must also be taken into account.

Perhaps you wonder how SR has survived all the scrutiny over the hundred years since 1905. The reason might be that Einstein's derivation is perfectly logical and correct as long as the world is four-dimensional. This means that mathematical minds analyzing it have not found anything wrong with it. Yet, it just does not make sense because it is not consistent. The scientist is confronted with the choice of either believing the math and Einstein and just accept the inconsistencies, or point out that something is not right with SR, as Dingle did, without knowing what it possibly could be. Since some people in science tend to rely more on mathematics than on common sense the SR theory has survived.

But, no longer!

Symmetry between Inertial Frames: Simultaneity

In the preceding discussion, I have considered relative motion between a stationary frame and a moving frame. Because of the relativity principle of SR, we should be able to pick any one of these two frames as being stationary. This is true; the inertial scale factor applies to *relative* velocity, which implies that any inertial frame may be considered stationary even when moving relative to some arbitrary reference frame. Therefore, the velocity, v, in our discussion may be taken as the relative velocity, and in any particular arbitrary frame that is chosen as reference we have $v = 0$.

Since the development presented above applies to any inertial frame as reference, it implies that each inertial frame "sees" the scale of space and time in other, moving frames as being contracted.

According to this scenario, a frame that appears to be flat and Minkowskian in a certain inertial reference frame will in another inertial frame, which is moving in relation to the reference frame, no longer be Minkowskian but will be "scaled Minkowskian." However, this perspective is reversed when exchanging the two frame designations so that the moving frame becomes the stationary reference.

At first, the proposition that *spacetime curvature is relative* might seem strange, but we are already familiar with this situation from GR, where we may choose our coordinates so that a local Minkowskian tangent frame always exists and where spacetime at other locations may appear curved in relation to this local frame. Therefore, the concept of inertial spacetime curvature is *relative*; each inertial frame is offering a similar perspective of moving frames as given by the inertial line-element.

This kind of symmetry may also be illustrated by a simile, letting the 4D Minkowskian spacetimes of inertial frames be represented by flat, 2D surfaces. Consider two different positions on a spherical surface, each with a local tangential plane. Although the relative perspectives from these two planes are the same, they are rotated relative to each other.

This is illustrated in figure 17, where the relative orientation is reflected by the angle Phi.

Note that this simile collapses the three spacetime dimensions into one spatial dimension.

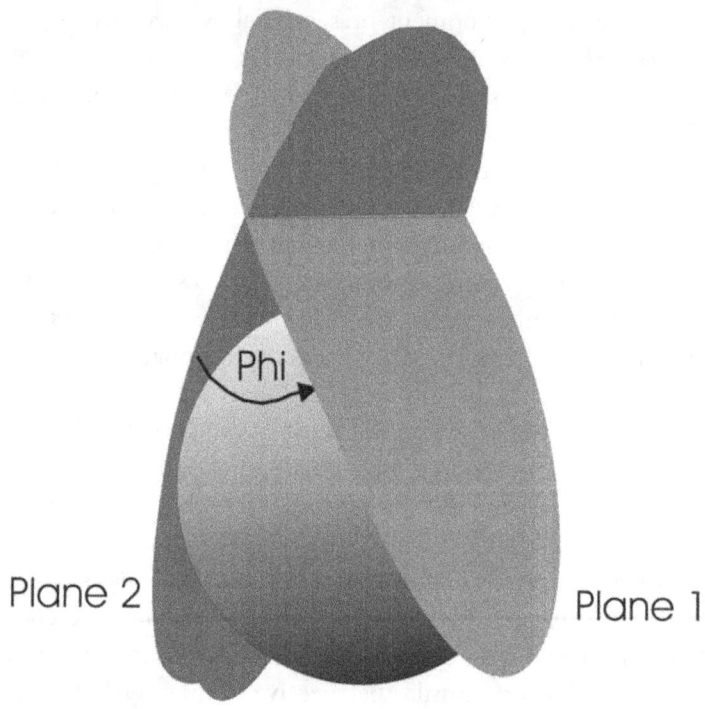

Figure 17: Illustrating an additional dimension

The two planar surfaces in the figure represent 4D spacetimes. Obviously, it is impossible to model this situation without taking into account the separating curvature, which corresponds to the dynamic scale. This is also true with

Minkowskian line-elements of inertial frames, which cannot be geometrically related by any transformation in 4D spacetime, like the Lorentz transformation attempts to do.

The figure above also nicely illustrates the limitation imposed by a missing dimension. In a plane's two dimensions, it takes at least two views to illustrate relative sizes, while only one view is needed in three dimensions. Similarly, two different perspectives are needed in SR to illustrate relative scales of inertial frames as experienced by inertial observers in two frames, while only one would be needed if we could visualize the additional scale dimension. This scale symmetry between inertial frames also preserves temporal symmetry.

The geometrical simile merely illustrates the concept of relative scale symmetry in three dimensions rather than in five and has no particular physical significance other than to help us visualize a symmetric dynamic spacetime scale. However, the figure shows how transitioning from one inertial frame to another reverses the relative scaling so that the perspectives from both frames remain the same. It is also clear that no 2D transformation constrained to a planar surface can describe the separation in three dimensions. Similarly, no 4D transformation exists that can describe moving coordinate frames separated in scale.

This interpretation is, of course, new, but it has the advantage of resolving a clear logical inconsistency of SR as well as explaining the nature of the inertial force.

A New Dimension

The scale symmetry between inertial frames implies another aspect of 4D scale transition. It suggests that line-elements that differ in scale belong to different spacetime manifolds of GR. GR deals with fixed 4D geometry in which there is no motion. Each inertial frame carries with it its own 4D spacetime geometry (Remember Zeno's Arrow-Paradox!) relative to which other inertial frames appear to have smaller relative scales.

Acceleration will cause transition from one 4D spacetime to another while their relative scales adjust. This may be further illustrated by the following thought experiment.

Consider three inertial frames: A, B, and C. Let the relative velocity between A and C be fixed and equal v. Frame B accelerates from A to C and will see changing relative scales of A and C. As seen from the accelerating frame B, frame A's scale decreases, while frame C's scale increases.

This kind of symmetry is not that unfamiliar to us. For example, due to visual perspective, an object seen at some distance away appears to be smaller and this perspective is symmetric, depending on distance as shown in figure 18.

Figure 18: Illustrating three dimensions by two different 2D perspectives

Similarly, the perspective of inertial frames is symmetric and only depends on the relative velocity.

If locally all inertial frames have the same Minkowskian geometry, but this geometry is contracted in a relative sense in other frames, we may say that we observe a moving frame merely as a projection onto the local frame; what we see is not what is experienced locally by a co-moving observer.

This may be visualized as illustrated in figure 19, where 4D spacetimes are represented by two 1D lines that are separated in two dimensions.

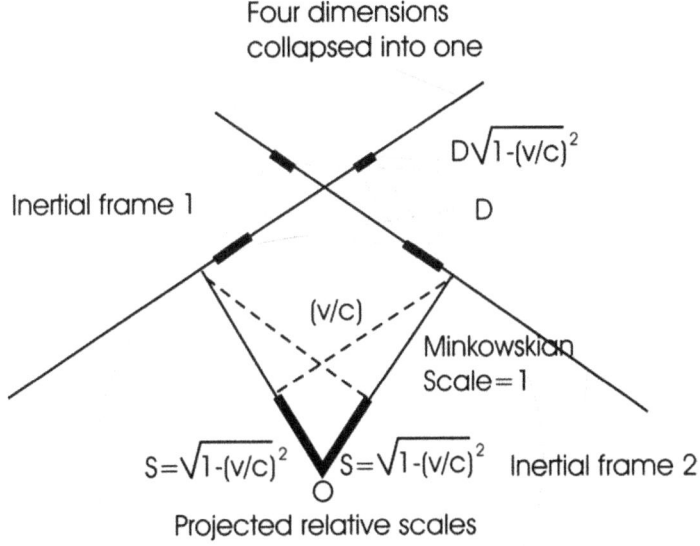

Figure 19: Illustrating projections of spacetime intervals

In this figure, length contraction and time dilation are shown as being caused by projecting one spacetime onto another. Acceleration would cause relative rotation of the two lines in the figure, and all projected distance increments approach zero when v approaches c, just as they do in SR.

Note that with the geometric interpretation, length contraction takes place *in all three* dimensions rather than merely in the direction of motion as is the case in SR.

A Relative Velocity Space

The possibility that the scale of spacetime is a relative concept may also be visualized by considering a 3D velocity space in which locations are defined by 3D velocity vectors. This space is relative in the sense that any inertially moving particle may be taken as being the reference located at the origin. Thus, the velocity at the origin in this frame is zero and any point in this space corresponds to a particular relative velocity. The corresponding 1D scalar space of relative scale-values may be derived from the 3D relative velocity space by forming the inertial scale-factor $S(v) = 1 - (v/c)^2$.

This representation allows us to visualize the dynamic nature of the relative scales of inertial systems and how all relative scale values will change simultaneously when moving from one inertial frame to another. It makes it easy to understand how the scale of spacetime can be relative in the sense that it only depends on relative velocities.

This insight is the key to understanding where SR went wrong: SR does not recognize the scale dimension.

Einstein on Inertia

I mentioned earlier that Einstein had hoped that GR would support Mach's Principle, according to which distant matter provides an absolute spatial reference frame. If GR could explain the existence of a cosmological reference frame, it might perhaps also explain Inertia. But he found that GR cannot do this; there must be something else that explains it, and Einstein seemed to have considered the existence of some kind of aether, albeit of a more nebulous form. Here is what he said [Einstein, 1924]:

> *"It is true that Mach tried to avoid having to accept as real something which is not observable by endeavoring to substitute in mechanics a mean acceleration with reference to the totality of the masses in the universe in place of acceleration with reference to absolute space. But inertial resistance opposed to relative acceleration of distant masses presupposes action at a distance; and as the modern physicist does not believe that he may accept this action at a distance, he comes back once more, if he follows Mach, to the ether, which has to serve as medium for the effects of inertia. But this conception of the ether to which we are led by Mach's way of thinking differs essentially from the ether as conceived by Newton, by Fresnel, and by Lorentz."*

This comment reveals that Einstein in 1924 was thinking about the possible existence of some unknown agent that causes inertia. Although GR changes the trajectories of freely falling particles by altering the properties of spacetime, this does not really explain why inertia exists even very far away from matter. This makes the presence of inertia mysterious and unexplainable. Something is obviously there even in the absence of matter, but what?

In the SEC theory, this "something" influencing the structure of spacetime at every location in the universe is the cosmological scale-expansion. Particles sense its presence via cosmic drag and perhaps, even more directly, via the oscillating scale of spacetime, which may sustain their existence. This oscillation, induced by the scale-expansion, also determines atomic time as an absolute cosmological temporal reference.

This would introduce a new aspect of the kind Einstein was looking for, but not as a property of the aether, at least not as usually imagined. The scale-expansion with its dynamic scale induces a cosmological background that acts locally to preserve the Minkowskian spacetime geometry. In short, particles sense when the speed of light changes during acceleration and adjust to the new situation by modifying their metrics. Note that this provides an onto-logical explanation to length contraction and time dilation; we might say that a particle does not "contort" via mechanical stresses, but instead it simply changes the local properties of spacetime to suit the conditions needed for its existence. Thus, physical properties of an accelerating particle do not change; instead, physical properties (geometry) of spacetime change to accommodate the particle.

As with the SEC expansion, spacetime may change dynamically "relative to itself," which becomes possible by the semi-discrete DIST process in which the 4D scale is a controlling force and all motions respond to the changing conditions of spacetime by adjusting the scale.

We might say that the cosmological scale-expansion plays the role of the aether; it is a reference that sets the stage for all motion.

A New Perception of Motion

This discussion suggests a new perception of motion, which involves not only space and time but also a dynamic spacetime scale. This new type of process cannot be modeled solely by positions in space and time as given by the classical equations of motion. We might ask where this could lead.

In the past, our understanding of motion has come under repeated scrutiny. As we already saw, the ancient Greek philosopher Zeno probed the limits of our understanding by his paradoxes, and recently we have found that energy on small scales is quantized and cannot be described by classical differential methods. However, the reason for this quantization has not been clear, and motion in the quantum realm is presently treated differently than in classical motion, which contributes to making GR incompatible with QT. However, this situation might, perhaps, be remedied by realizing that inertial frames might have different relative metrical scales.

The dynamic incremental scale transition (DIST) process implies a new and very different aspect of motion. It models the dynamic scale of spacetime without altering the spacetime geometry in relation to an observer participating in the motion. Relative to an inhabitant of the universe as well as any inertial observer, the geometry remains the same (on the average), but all four metrics change incrementally.

Thus, all accelerating motion implies changing scale in a relative sense; an observer in an inertial reference frame sees the scale change in another, accelerating, frame. But, although the scale of spacetime may change in this relative sense, it does not change locally in relation to an observer participating in an accelerating motion.

Spacetimes of different accelerating observers are therefore not the same; they differ in a fifth dimension—the relative scale of spacetime!

Ehrenfest's Paradox

Paul Ehrenfest was one of Einstein's closest friends who posed a paradox known as Ehrenfest's Paradox. Here is an extract from *Wikipedia; Ehrenfest's Paradox*:

> "*In its original formulation as presented by Paul Ehrenfest 1909 in the Physikalische Zeitschrift, it discusses an ideally rigid cylinder that is made to rotate about its axis of symmetry. The radius R as seen in the laboratory frame is always perpendicular to its motion and should therefore be equal to its value R_0 when stationary. However, the circumference ($2\pi R$) should appear Lorentz-contracted to a smaller value than at rest, by the usual factor $1/\gamma$. This leads to the contradiction that $R = R_0$ and $R < R_0$.*
> *(Note that a cylinder was considered in order to circumvent the possibility of a disc "dishing" out of its plane of rotation and trivially satisfying $C < 2\pi R$. Subsequently, when a rotating disc is substituted, it is assumed that this distortion possibility is also excluded).*
>
> *The paradox has been deepened further by later reasoning that since measuring rods aligned along the periphery and moving with it should appear contracted, more would fit around the circumference, which would thus measure greater than $2\pi R$.*
>
> *The Ehrenfest paradox may be the most basic phenomenon in relativity that has a long history marked by controversy and which still gets different interpretations published in peer-reviewed journals.*"

Let's analyze this paradox in light of our new understanding. First, we note that the contraction is not real but merely expresses a projection from the moving frame onto the stationary frame. The circumference actually remains the same, a fact that may be realized by the following argument.

Consider a point on the moving periphery. From the preceding discussion, we conclude that due to the DIST process with its incrementally adjusting scale, spacetime at this point will always at each instant remain locally Minkowskian. Therefore, an observer on the periphery will locally measure the same increments along the periphery of the rotating cylinder as if the cylinder did not rotate. Since this is true at any point on the periphery, the circumference measured by the moving observer must be the same as when the cylinder does not rotate. We conclude that the smaller circumference is a relative, apparent phenomenon.

You might perhaps wonder how the spacetime curvature creating the centrifugal force is taken into account in this scenario. It's done via the dynamic spacetime scale, which during each step of the DIST cycle models the centrifugal force as a gravitational-type phenomenon. During rotation the metrics of spacetime adjust to the new direction of the velocity, which creates a metrical gradient and an inertial force. The metrics are "reset" incrementally by the DIST process in order to locally preserve the Minkowskian line element. This means that relative to a co-moving observer, spacetime always remains the same in inertial frames as well as in accelerating frames.

In other words, locally the geometry is preserved during motion. However, in relation to a stationary observer, the metrics change with motion.

This seems to conflict with the modeling of gravitation by coordinate metrics that depend on location in space and time. However, GR cannot model the progression of time and therefore cannot model motion as it is perceived by an observer participating in the motion. In the SEC, the progression of time corresponds to the transition between 4D spacetimes. Likewise, motion in space involves the transition between consecutive 4D spacetimes embedded in a five-dimensional hyperspace with the scale as the fifth dimension.

Summarizing this Chapter

Dynamic spacetime metrics modeled by the DIST process might resolve the ancient puzzle of inertia. It could be a curved spacetime phenomenon similar to gravitation. As we saw in the previous chapters, the DIST process also offers an improved cosmological model.

By this ontological explanation to inertia, the scale of spacetime depends on relative velocity and diminishes with increasing velocity as given by the scale-factor $S(v) = 1 - (v/c)^2$. Coordinate increments in a moving frame merely appears as "projections" onto the local 4D spacetime. Consequently, projected increments no longer agree with the coordinates experienced locally by a moving observer. If this difference is not recognized and acknowledged, it will give rise to inconsistencies that become acute in the Twin Paradox.

Note that the projected coordinates of the moving frame actually are "real" in the sense that they are what a stationary observer actually experiences when observing a moving frame, which explains the success of SR. However, in their local inertial frames, all co-moving observers experience the same coordinate metrics. Consequently, all observers will agree on elapsed time intervals when their clocks are compared side by side, regardless of their different motion histories.

The relativistic time of SR disappears in favor of an absolute temporal cosmological reference.

Since by the SEC theory, cosmic drag diminishes relative velocities of freely moving objects, a spatial reference frame is induced as the frame toward which all free motion converges. Together with the temporal reference given by atomic time, this reestablishes Newton's "absolute space" as a firm and consistent stage for all existence. This reference is induced by the cosmological scale-expansion.

If these conclusions hold up, SR will have to be revised by accommodating a changing 4D scale or by adding a fifth dimension. In retrospect, Einstein was right with his two SR postulates as well as in assuming that the geometries of all inertial frames are locally Minkowskian. However, he appears to have made a mistake by assuming that coordinates of moving frames may be connected by the LT *while preserving their metrics in a relative sense*. This mistake would not be surprising since the GR theory was not yet available when SR was introduced in 1905, which made it unlikely to consider a changing 4D scale. SR has in effect covered up the changing spacetime scale and prevented the discovery of the origin of inertia.

If the reader has had the tenacity to stay with me up to this point, the contours of a new and very different worldview should now have started to emerge out of a fog of confusing misconceptions.

CHAPTER 8

Quantum Theory and Its Link to General Relativity

There are many comments expressing confusion with quantum theory, and I could mention a famous quote by Albert Einstein, *"God does not play dice!"* Instead I will use this one from a letter to Heinrich Zangger-May 20, 1912 (Albert Einstein Archives 39-655):

> THE MORE SUCCESS THE QUANTUM THEORY HAS, THE SILLIER IT LOOKS.
>
> —ALBERT EINSTEIN

I also like another one by Richard Feynman in his book *"QED, The Strange Theory of Light and Matter"*, Princeton University Press, (1985):

> I HAVE POINTED OUT THESE THINGS BECAUSE THE MORE YOU SEE HOW STRANGELY NATURE BEHAVES, THE HARDER IT IS TO MAKE A MODEL THAT EXPLAINS HOW EVEN THE SIMPLEST PHENOMENA ACTUALLY WORK. SO THEORETICAL PHYSICS HAS GIVEN UP ON THAT.
>
> —RICHARD FEYNMAN

Not many would admit to not understanding quantum mechanics, but I guess Feynman's stature as a famous Nobel laureate allowed him to do it. It speaks volumes about our current state of knowledge; we simply do not understand our quantum world.

As we shall see, Quantum Theory (QT) might be explained by the additional scale-dimension and the DIST process, a fact that perhaps may be seen as a direct confirmation of the new physics presented in this book. The generalization of GR by the DIST process allows the derivation of QT from GR! This would bridge the gap between GR and QT and thus resolve perhaps the greatest mystery of modern physics.

If you like classical physics, you might not like QT; at least I didn't enjoy it when first encountering it as a student. QT, as it usually is taught, does not give any explanation to the underlying ontology, which I believe exists. In my mind QT is not physics but mathematics; it is a number of mathematical rules without ontological explanation. QT seems to work very well, but nobody knows how or why.

To begin with, nobody knows why Nature is quantum mechanical—its "nuts and bolts" remain hidden. Practicing QT is similar to using a computer without knowing its interior function; you push the right buttons and out comes the desired answer. But nobody really knows what is inside the black computer box of QT. Generations of students who have struggled with QT should take heart; they may take comfort from learning that their teachers really don't know what's going on in the black box either! In other words, QT is like ancient Chinese medicine; the remedies work, but there is no explanation to why or how they do it. Here is another quote by Feynman:

"I think it is safe to say that no one understands Quantum Mechanics."

The material in this section was published in [Masreliez, 2005a] and its mathematical details may be found in Appendix IV.

Current QT Epistemology

Most people interested in physics are familiar with the historical development of QT, but let me quickly give you my own impressions. The reader acquainted with this background may prefer to go directly to the next section.

One of the most perplexing aspects of the quantum world is the so-called wave-particle duality. This is the mysterious property where in some experiments a particle behaves like a wave and in other experiments like a particle. If we are looking for a particle, we will see a particle, and if we are looking for a wave—for example, in an interference experiment, we see a wave. This greatly bothered Niels Bohr, who eventually "solved" the problem by his *Principle of Complementarity*. He postulated that we simply must accept that a particle sometimes behaves like a particle and sometimes like a wave of unknown ontology, but with probabilistic interpretation. In his Copenhagen school, this became a mantra, and all efforts to find an ontological explanation of QT were discouraged. The consensus was that the missing ontological explanation to QT should simply be accepted; we should accept that the workings of the universe at atomic and subatomic levels will forever remain hidden.

However, several prominent people, including Einstein, Schrödinger, and deBroglie, were not convinced or happy with this argument and rejected Bohr's Principle of Complementarity. They felt that something important must be missing and that complementarity was a "copout" that excused us from trying to find the deeper truth. The ensuing debate between Einstein and Bohr is historic. Unfortunately, today most people think that Bohr won by overwhelming the opposition, although recently this consensus seems to be changing.

Bohr had successfully attracted several leading young scientists to his Copenhagen school, which was one of the few surviving centers of learning in Europe after the devastation of World War II. With his forceful and tenacious personality, he created a strong following committed to a certain interpretation of QT, which became known as the *Copenhagen School*. However, as we shall see, this might have been a step in the wrong direction.

Einstein thought that the QT theory was incomplete and expressed his unhappiness in a letter to Max Born:

"Quantum mechanics is certainly imposing but an inner voice tells me it is not yet the real thing."

The Einstein–Bohr Debate

In a series of exchanges, Albert Einstein and Niels Bohr debated the intricacies of QT. These debates are of interest not only for their historic value but also because of the two protagonists' different personalities. The following is my personal take on this part of history.

Einstein was a man who deeply contemplated the world with an overriding desire to understand it. He wanted to learn the mind of the Creator. Niels Bohr, on the other hand, was the leader of the so-called Copenhagen School of QT. Bohr's main interest seems to have been to try to make sense of our quantum world and, if possible, establish a seamless account that could form the basis for the new quantum epistemology. Bohr had a very intense and tenacious personality while Einstein assumed a rather laid-back and withdrawn posture.

Their intense but friendly exchanges expressed these different personalities and objectives. Usually these took the form of Einstein challenging one or another aspect of the Copenhagen epistemology that Bohr was trying to establish. Bohr responded to these challenges and defended his views, but repeatedly had to reevaluate his position in response to Einstein's challenges. No doubt, QT greatly gained from this debate.

Bohr had, over a long time, tried to find an explanation to the puzzling fact that a particle sometimes behaves like a particle and sometimes behaves like a wave depending on the experimental setup. In spite of having discussed this strange behavior at length with several leading personalities in physics of the time, he had not been able to find any explanation to this enigmatic behavior. For him this was probably an unacceptable situation because it would leave a glaring gap in the presentation of quantum theory. In desperation, he tried to resolve the issue by his Principle of Complementarity. He argued that we simply must accept that nature has two faces; that a particle both can act as a particle and a wave. The aspect turning up in a particular experiment depends on the experimental setup; if we are looking for a particle, we will see a particle, and if we are looking for a wave, we will see a wave. According to the Principle of Complementarity, the particle and the wave should be seen as complementary aspects, but why this should be the case cannot be explained. We should simply accept it.

Einstein did not agree and insistently argued that something important must be missing in QT; he felt that something definitely remained to be discovered. However, Bohr would not accept this, probably because it would leave his epistemology incomplete. He discouraged all attempts to explain the wave-particle duality—for example, he rejected David Bohm's (1952) alternate theory with its pilot-wave.

In retrospect, we find that Einstein was right; something important *is* missing in the Copenhagen interpretation. However, what is missing cannot be found without violating Einstein's own theories.

Ironically, Einstein's two relativity theories had by their very success inadvertently blocked access to the path leading to the explanation of the wave-particle duality as well as to an ontological explanation to the quantum world.

Einstein's two theories constrain the world to the four dimensions of spacetime. It ignores the dynamic spacetime scale, which acts beyond these four dimensions and is the domain of quantum influences. As we shall see, the wave-particle duality may be explained if particles are standing wave oscillations in the scale of spacetime at the Compton frequency. When they move, they will then induce the de Broglie matter-wave as phase modulation of the Compton wave.

Compton oscillation and the de Broglie matter-wave

A particle, including a photon, is associated with Compton oscillation. Its energy is given by $E = hf$ where h is Planck's constant and f is the Compton frequency. This relationship between energy and Compton frequency holds true for all particles and is for particles with rest mass given by $mc^2 = hf$. This means that even at rest particles oscillate at the Compton frequency.

The deBroglie matter-wave is associated with particles in motion. For photons its wavelength coincides with the wavelength of light. However, for particles moving at lower velocities the deBroglie wavelengths are longer and at rest they become infinitely long.

According to current understanding both these waves are quantum mechanical in nature. However, here they will be interpreted as being modulations of the scale of spacetime.

Therefore, the wave nature of particles might merely be a simple and direct consequence of their metrical oscillatory nature. In this context it should also be noted that the deBroglie matter-wave is due to metrical scale adjustment associated with particles in relative motion. Thus, the matter-wave is also a consequence of the dynamical scale of spacetime.

An Ontological Hint

Let us take a second look at the cyclic DIST process that illustrates the cosmological scale-expansion cycle. This might very well be a simplified picture, since it is conceivable that the cosmological expansion simultaneously could take place via several loops at different frequencies. However, the DIST loop indicates that *the metrical scale of everything in the cosmos oscillates relative to a co-expanding observer*, which suggests that there might be some kind of connection between the cosmological scale-expansion and QT, since QT is dealing with waves and discrete processes. I will argue that this actually is the case:

The oscillating scale-expansion explains our quantum world.

By the cosmological DIST cycle, the scale expands by a tiny fraction during each cycle. At the end of each DIST loop, spacetime "jumps into" a new, slightly larger scale through a discrete scale change. As experienced by an inhabitant of the universe who expands together with spacetime, this jump in effect resets the relative scale of spacetime, and therefore, it appears to oscillate at extremely high frequencies. This oscillation cannot be directly observed because the amplitude, superposed on the scale, is extremely small and the frequency extremely high, but I believe that this is what causes the familiar vacuum fluctuations, and that it is the domain where QT is active. If you wonder why Nature is quantum mechanical, the answer could be that it has to be, because of the Rule of Timeless Existence. The RTE implies discrete scale-expansion and therefore also QT!

The de Broglie Matter-Wave

Now, let us consider a small spatial region (a particle) with a metrical scale oscillating at a very high frequency. We can model this in GR by a Minkowskian line element with oscillating metrics. To model motion, we apply the Lorentz transformation (LT) or the Voigt transformation (VT). We then find that motion will cause spatial modulation of the phase of the scale oscillation. This is a relativistic consequence of the term xv/c^2 in the temporal transformation. If the scale oscillation, which is associated with the particle, matches the so-called Compton frequency, *we recognize this phase modulation as being identical to the de Broglie matter-wave!*

The Compton frequency, f, is related to the energy, E, of a particle by the relationship

$$E = h \cdot f.$$

Here h is Planck's constant. This simple but important observation suggests that:

- Scale oscillation of small amplitudes at the Compton frequency might accompany and sustain all particles.

- The deBroglie matter-wave is a modulation of the Compton oscillation of the scale. It is a relativistic effect, which is a direct consequence of a particle's motion.

- Thus, the wave and particle aspects of QT are inseparable—they are two sides of the same coin.

- This would immediately explain the wave-particle duality. These simple but important observations reveal a great deal about the nature of QT.

This suggests an ontological explanation: The QT wavefunction represents modulation of a particle's Compton oscillation in the scale of spacetime. By this interpretation, the quantum wave is not a separate, independent entity but represents modulation of already existing particle-oscillation. Like a radio signal is modulated to transmit speech and music, the Compton oscillation is modulated by the QT wave-function. The complex nature of the wavefunction now finds its explanation; it expresses amplitude and phase modulation of the Compton "carrier" wave. This suggests that the essence of QT is oscillation in the scale of spacetime and that the QT wave-functions are real physical entities, not just probabilistic functions. The modulation of spacetime is primary; Born's probability interpretation is secondary.

This would also discredit the Copenhagen interpretation with its Complementarity Principle by providing a simple and direct physical explanation to QT. People who over the years have felt that something is missing in the way QT is to be understood would be right. Thus, Einstein might have been right and Bohr wrong after all! However, ironically this should not be blamed solely on Bohr and the Copenhagen School because it could also be due to shortcomings of GR, which cannot model the DIST process. Both theories will have to be modified before reconciliation becomes possible.

According to the Copenhagen School, the wave-functions represent particles without giving any ontological explanation. Motion of a particle is modeled based on the corresponding wave-function. By this approach, QT wave-functions are interpreted as being the primary observable entities while the particles become secondary. Thus, quantum mechanics deals with wave-functions rather than with particles. This is like learning the properties of an object from the behavior of its shadow. Like the shadow, the wave-function depends on the surrounding geometry and might give strange interpretations if one thinks the shadow is actually the object, in particular if the object casts several shadows, which in QT would correspond to different branches of the wave-function.

The new interpretation suggests that particles could be standing wave oscillations in the metrics of spacetime sustained by the cosmological scale-expansion. Oscillation of the metrics can generate both positive and negative energy in GR, and it is possible that the Compton oscillation generates a particle's rest mass energy. In this case, matter (particles) would be nothing but oscillating spacetime energy.

Let's see if this rather speculative conjecture finds additional support!

The Origin of Mass

The currently accepted model of particle physics proposes that all particle masses are induced via a hypothetical particle called the Higgs boson. Nobody knows if it exists, but it is now being searched for by CERN's 8 billion dollar Large Hadron Collider. As of October 2011 it has not been found.

The origin of mass and energy is somewhat of a mystery in the SCM and the origin of the cosmological Dark Energy is unknown. However, in the SEC this energy is steadily being induced by the expanding scale of space and time. It may easily be shown that an oscillating scale will induce spacetime energy with an energy-momentum tensor of the same form as the Cosmic Energy Tensor of the SEC theory. Like the Cosmic Energy Tensor it has cancelling positive and negative contributions. This might be the Zero Point Energy of vacuum.

However, it is possible that all four spacetime metrics do not oscillate in exactly the same way but exhibit internal phase differences or overtones.

This could be the origin of mass energy.

This means that the cosmological scale-expansion could be the ultimate energy source, not only for the Dark Energy, but also for all matter-energy. And, as already mentioned, this matter-energy is exactly balanced by the negative energy of the corresponding gravitational field! The net energy of the universe disappears. However, exactly how the fundamental particles are formed from oscillating spacetime metrics is not yet known.

The de Broglie–Bohm Pilot-wave

Over the years since the discovery of the matter-wave, several attempts have been made to find an ontological interpretation for QT. Louis de Broglie suggested at the Solvay conference in 1927 that a particle might be guided by a *pilot-wave* directly related to the QT wave-function.

At this meeting Wolfgang Pauli challenged him to explain what happens to his pilot-wave at the "scattering" of a particle, that is, when a particle hits an object and scatters away. This is usually modeled as a single QT wave-function that splits up into a superposition of several different components representing different scattering outcomes. A single pilot-wave corresponding to this superposed wave-function cannot explain the different possible trajectories taken by the scattered particle, since it would mean that the particle had to follow several different trajectories at the same time.

Later, in the 1950s, David Bohm independently re-vived de Broglie's idea [Bohm, 1952]; [Bohm and Vigier, 1954]. He attempted to counter this scattering challenge by speculating that "decoherence" quickly occurs between the different branches of the scattered wave-function and that the scattered particle selects only one of the possible branches, leaving the other branches empty. However, he did not clarify the reason for this decoherence.

Bohm's explanation should be compared to how QT is being taught today. The different branches of the wave-function are thought to represent "potentialities"; these branches represent different possibilities that the particle will follow a particular branch. After scattering, but before an observation is made, the particle is believed to be "hov-ering" in all different branches simultaneously; the act of observation "collapses" the wave-function into one of the possible branches. This "collapse of the wave-function" is a very strange mental "move," which has been discussed and debated at length over the years. It is central to current QT epistemology and has been the subject of many articles and much speculation. It is undoubtedly the most unsatisfactory aspect of the Copenhagen interpretation; it is something unexplainable and mysterious.

Let me suggest another interpretation.

The Compton carrier frequency is proportional to the relativistic energy of the particle. This energy, which also includes its kinetic energy, changes with the particle's velocity, which means that the relativistic Compton frequency also changes with the velocity. In scattering, the particle bounces off in a different direction and its velocity will change. The Compton carrier frequency then shifts slightly and may therefore select a different branch of the QT wave-function. By this mechanism, the different branches of the wave-function will become decorrelated, just as Bohm guessed, *due to their different Compton frequencies*. Just like radio signals in different bands do not interfere, the branches of the wave-function do not interfere because their carrier waves differ. As soon as we realize that the QT wave-functions do not have an independent existence but merely modulate the Compton carrier wave, we begin to understand what is happening. Possible trajectories appear as different branches, one for each Compton frequency. After scattering, the particle will take one of these possible trajectories corresponding to its energy. This eliminates the troublesome and conceptually ugly "collapse of the wave-function." The selection of a particular branch simply corresponds to a particular scattering velocity.

To my knowledge his explanation has not been suggested in the past because quantum mechanics deals with the QT wave-functions, not realizing that they represent modulation of the metrical Compton wave.

The de Broglie–Bohm Pilot-wave is the Geodesic of GR

There are more recent versions of Bohm's theory championed by John Bell, [Bell, 1987] and others—for example, Peter Holland [Holland, 1993)] and Dürr, Goldstein, and Zanghi [Dürr, Goldstein, Zanghi, 1996]. They show that a consistent quantum mechanical theory may be derived based on just three assumptions:

There exists a function, ψ (of unspecified ontology) with the following properties:

1. It satisfies Schrödinger's wave equation.

2. The momentum p of a particle satisfies the pilot-wave relation:

    $p = \hbar \cdot \mathrm{Im} \dfrac{\nabla \psi}{\psi}; \nabla =$ gradient operator (Here Im stands for the imaginary part).

3. Some random disturbance is present.

David Bohm and his followers have shown that the pilot-wave relation together with the Schrödinger equation may be used to construct a theory that in all respects is equivalent to elementary quantum mechanics, provided that also some random disturbance is present.

However, one puzzling aspect of Bohm's theory is the nonlocal character of the pilot-wave. Since it contains the ratio between two functions, the momentum p could become very large even when the magnitude of the wavefunction is close to zero. Therefore, it could exert influence over vast distances even at very low amplitudes. It is difficult to understand how this might be possible and how distant wave-functions of negligible power could influence the local motion of particles. Bohm called this property *"active information"*, proposing that the pilot function somehow "informs" each particle how to move without exerting any physical force. Since this appears rather speculative, the mysterious long-range action could have discouraged more substantial support for Bohm's theory. It might also have deterred Einstein from fully supporting the de Broglie interpretation. But there is a physical explanation to the pilot-wave that Einstein probably would have appreciated.

If the metrics oscillate, de Broglie–Bohm pilot-wave relation may be derived from the geodesic equation of GR!

At first I thought that this only holds for small velocities [Masreliez, 2005]), but a closer investigation has shown that it is true in general. Oscillating metrics in GR would not only explain why there is a quantum world but also explain the role of the pilot-wave. In the derivation of the geodesic from GR (assuming that the metrics oscillate), we find that the sum of a few oscillating terms must equal zero. This leads to a geodesic equation that involves *velocity* rather than acceleration like in the usual geodesic.

This is the pilot-wave relation.

It is derived in Appendix IV. With this interpretation, the pilot-wave finds its natural explanation; it expresses how a particle responds to modulation of the spacetime metrics. A particle moves on its GR geodesic without being subjected to any external force. A particle's trajectory could be curved, and certain regions of resonance might be preferred. In this way, the pilot function could influence the motion of a particle without energy transfer even non-locally via the scale of spacetime. A particle follows a path in spacetime determined by its oscillating metrics. The wave-function that modulates the metrics shifts the phase depending on the trajectory and the surrounding geometry. Regions of resonance are created where the phases of different alternate paths coincide. The particle prefers these resonating regions, which means that energies and locations might be quantized. Resonances will occur only at certain Compton frequencies (energies) and at specific spatial locations. This explains the discrete nature of QT and its wave-mechanical features.

The Schrödinger Equation

I mentioned that Bohm and his followers were able to show that classical QT may be derived from the pilot-wave if the wave-function ψ *satisfies the Schrödinger equation* and there also is some random disturbance. It turns out that the Schrödinger equation also may be derived from GR with oscillating metrics if we assume that the so-called "Ricci scalar" of GR equals zero. The Ricci scalar also disappears for the gravitational field in vacuum (disregarding the small contribution from the SEC expansion), and it seems reasonable that this also should be the case for oscillating metrics.

This assumption leads to a wave equation from which the Schrödinger equation may be derived, assuming that the phase of the metrical Compton oscillation depends on a field potential V, which is a function of location. This is a natural assumption if the field potential changes the velocity, because it will then also change the frequency and the phase of the Compton oscillation. The Ricci scalar turns out to be a complicated sum of wave terms. We get the Schrödinger equation by setting the sum of these terms equal to zero; see [Masreliez, 2005a]. Appendix IV gives this derivation.

Thus, modeling scale oscillation in GR yields the Schrödinger equation.

This tells us that if a field potential V influences the phase (velocity) of the oscillating spacetime metrics, a particle's Compton carrier wave is modulated by wave-functions that satisfy the Schrödinger equation. This makes perfect sense, because the phase of the metrical oscillation responds to the particle's energy, which depends on the potential V. It is interesting that this derivation of the Schrödinger equation holds true independently of the particle's trajectory, which means that the Schrödinger equation expresses spacetime resonances that only depend on the particles energy (Compton frequency), on the imposed field V, and on the geometry; the equation does not model a particle's motion. Like a terrain map, the Schrödinger equation describes peaks and valleys and does not describe motion through this terrain. This also is consistent with the continuous representation in terms of wave-functions, which do not model quantum jumps.

However, there is one new aspect, which is crucial for understanding QT. Each wave-function solution to the Schrödinger equation typically depends on the spatial coordinates, and describes the modulation of phase and amplitude of the Compton carrier. The Schrödinger equation describes how the oscillating metrics *would be modulated if a particle with certain energy were present at a certain location.*

In my paper [Masreliez, 2005a], I also derive the Schrödinger equation for the electromagnetic field from oscillating metrics in GR. We might therefore speculate that the electromagnetic field could be a modality of spacetime metrical oscillation for which the mathematical CURL of a vector field (the electromagnetic vector potential) does not disappear. This would explain the many similarities between the electromagnetic and QT waves such as interference.

Since both the pilot-wave and the Schrödinger equation may be derived from GR if the metrics of spacetime oscillate, quantum mechanics follows directly from GR since random disturbance is always present. This provides a clear and direct link between GR and QT that previously has been missing!

The Double-Slit Experiment

Next, I will make a few comments on the double-slit experiment, which is often used as an introduction to the quantum world. Many, including Feynman, have given up trying to make sense of this seemingly mysterious experiment. But there may be a simple explanation. Here is the experiment.

Consider a particle moving toward a screen with two narrow slits. After passing through one of the slits, the particle strikes a second screen, where an interference pattern develops *even when particles arrive one at a time.* When I say "interference pattern," I mean that a particle seems to prefer certain fringelike bands on the screen, which gradually will appear after many particles have passed through the slits. This unexplainable and strange phenomenon initially motivated the particle/wave duality idea and Bohr's Principle of Complementarity. The interference indicates the presence of some kind of wave while the dots where particles hit the screen show that there are individual particles.

According to the standard interpretation, a particle somehow simultaneously passes through both slits and strangely "interferes with itself." The wave-function with its different branches corresponding to the fringes on the screen collapses when the particle strikes the screen.

David Bohm and others have shown that an interference pattern develops if his pilot-wave guides the particle, assuming that the pilot-wave simultaneously passes through both slits. However, this does not really explain the physical mechanism at work either. This problem is addressed here, together with a possible explanation.

Explaining the Double-Slit Experiment

When the particle passes through one of the two slits, it might randomly become slightly deflected—for example, by interacting with an edge.

After passing through one of the two slits, the particle's matter-wave interferes with the double-slit geometry.

This sets up a wave pattern behind the screen in the scale of the oscillating spacetime, which depends on the particle's location and velocity, and guides the particle via the corresponding geodesic. *Thus, the particle is guided by its own matter-wave and by the double-slit geometry.* The particle prefers regions with large wave amplitude and avoids regions with small amplitude. It is likely to end up in one of several interference fringes. Should it initially by chance move into a region with small interference amplitude, where the wave-function is close to zero, the geodesic will guide it into a region with larger amplitude. Remember that the momentum becomes large when ψ is small, which means that the particle avoids regions with small ψ.

Figure 20 shows a numerical prediction of how the particle fringes could be created if the spacetime metrics oscillate. This prediction is based on analytic expressions for the geodesic derived in [Masreliez, 2005a].

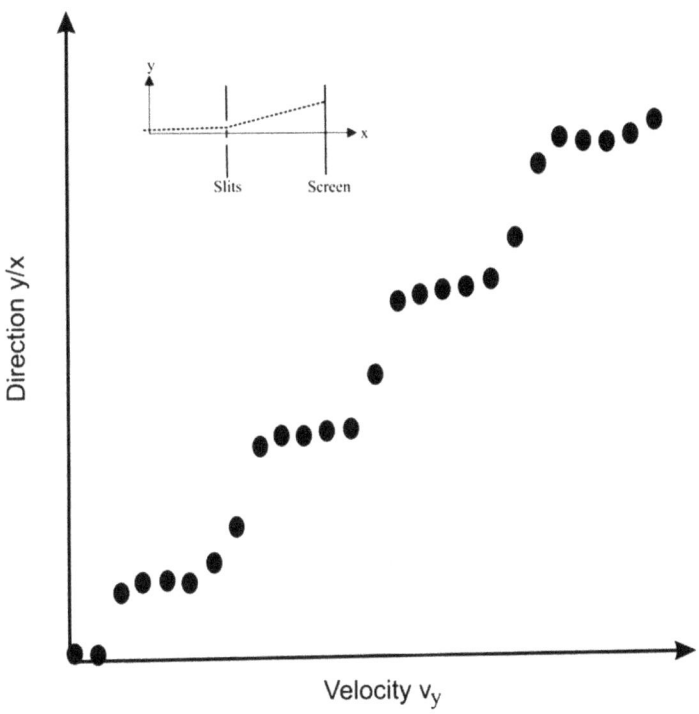

Figure 20: Computed double slit fringes

Thus, a particle may be guided by its own matter-wave, with the new and interesting insight that the guiding mechanism is the geodesic of GR. This connects QT firmly with GR.

It also illustrates a unique property: the guiding action is controlled via feedback. Spacetime resonance guides the particle, and the resonance pattern depends on the particle's motion. This explains how a particle finds its way into interference fringes and how resonance patterns surrounding the atomic nucleus constrain electrons to their orbits.

We saw that wave-functions are solutions to the Schrödinger equation,. They describe the response of spacetime *if a particle were to be present at the location given by the wave-function coordinates*. Thus, these wave-functions are not "active" unless a particle is present. This clarifies the role of the wave-functions and explains how they represent potentialities rather than physical waves. The quantum wave that modulates the metrics will materialize only when activated by the presence of a particle.

The Compton carrier also explains why a multi-particle wave-function only depends on particle locations and not on their velocities. Different velocities mean different relativistic Compton frequencies and therefore non-interference. Interference will only occur if identical particles move at the same velocity, and in this case, interference only depends on the locations of the particles. Thus, wave-functions modeling interference imply that particles are in the same energy state with the same Compton frequency. Their velocities do not appear explicitly because they are the same for all particles. If this is not the case, there is no interference.

Finally, let me offer a possible physical explanation to why particles might prefer locations with positive interference where the wave amplitude is large. If a particle is sustained by the Compton oscillation, its energy must somehow originate in the cosmological scale-expansion. In resonating states, the energy needed to sustain the particle is less than in other states. Since in nature energy is minimized, resonating states are preferred.

Spooky Action at a Distance and Quantum Entanglement

QT implies the existence of seemingly instantaneous influences between particles well separated in space, which would violate the claim that that highest possible velocity is the speed of light. Einstein rightly believed that either something must be wrong with this or that something is lacking in our understanding. He called it *spooky action at a distance*. He opposed QT throughout his life, contending that something important must be missing in the theory, and concluded that QT is an incomplete theory.

Here is an excerpt from a paper he published in 1948 [Einstein 1948], where he expresses his concerns:

"If one asks what, irrespective of quantum mechanics, is characteristic in the world of ideas in physics, one is first of all struck by the following:

The concepts of physics relate to a real outside world, that is, ideas are established relating to things such as bodies, fields, etc., which claim "real existence" that is independent of the perceiving subject…. It is further characteristic of these physical objects that they are thought of as arranged in a spacetime continuum. An essential aspect of this arrangement of things in physics is that they lay claim, at a certain time, to an existence independent of one another, provided these objects are situated in different "parts of space". Unless one makes

> *this kind of assumption about the independence of existence (the "being thu") of objects which are far apart from one another in space, which stems in the first place from everyday thinking, physical thinking in the familiar sense would be impossible. It is also hard to see any way of formulating and testing the laws of physics unless one makes a clear distinction of this kind."*

Here is another quote from the same source:

> *"There seems to me no doubt that those physicists who regard the descriptive method of quantum mechanics as definite in principle would...drop the requirement... for the independent existence of physical reality present in different parts of space; they would be justified in pointing out that quantum theory nowhere makes explicit use of this requirement. I admit this, but would point out: when I consider the physical phenomena known to me...I still cannot find any fact anywhere which would make it appear likely that {the requirement} will have to be abandoned."*

These statements express Einstein's doubts regarding QT in clear language. He states his unwillingness to accept non-locality without any ontological explanation for it. It appears that he wanted to be able to understand QT at a deeper level and not merely to accept spooky non-locality as being an unexplainable fact. He also makes the point that non-locality would conflict with SR.

Clearly Einstein's reasoning makes perfect sense provided one doesn't know the existence of a fifth dimension in the form of the dynamic scale in addition to the four spacetime dimensions. We should sympathize with him rather than discard his objections as an old man's ruminations, because nobody could claim to understand QT, not even Richard Feynman. And we should admire Einstein's uncompromising conclusion that the Copenhagen interpretation cannot be the last word. In fact, QT cannot be understood in the same way as classical physics is understood. We can visualize classical physics, but we cannot visualize QT within the 4D world of SR and GR.

In the past, nobody suspected that the four dimensions of spacetime were insufficient to describe the world. They did not realize that *the scale of spacetime may participate in all dynamic processes as a hidden fifth dimension.* And, since there is no speed of light constraint for influences via the metrical scale, which may act instantaneously by curving space, the troublesome non-locality of QT might find a physical explanation. What seems to be "spooky action at a distance" could be nothing but how a particle responds to the metrical wave-field. If this response does not transmit energy but acts via a geodesic, it could be instantaneous and could alter conditions even at a very distant location, which might change the outcome of a measurement there.

By this mechanism, the outcome of a measurement taken of a certain particle "A" may influence the outcome of a distant measurement of another particle "B." We say that A and B are "entangled." This entanglement could take place via the scale of spacetime beyond its four dimensions. This makes sense since we saw that the QT wave-functions could be modulations of the scale of spacetime.

QT cannot be explained without knowing about this dependence via the scale, and if we don't know about it, we might, to use Einstein's words, believe that there is spooky action at a distance.

But, by incorporating the new scale dimension, QT becomes explainable.

Hence, influence via the scale would explain quantum entanglement by which the local behavior of a particle may depend on another possibly very distant particle. This entanglement acts instantaneously in violation of any velocity constraint like the speed of light. We might visualize this as a scale resonance condition in a fifth dimension between the particles that makes them act in unity.

This would also imply simultaneity, which is in conflict with SR. However, the new theory of Inertia allows simultaneity since it is compatible with absolute time.

The SEC theory implies non-local action via the changing cosmological scale; it assumes that the scale-expansion acts simultaneously across the universe. If this wasn't the case, regions with different scale-factors could coexist, which would create streaming galaxy motion due to gravitational gradients. Although such streaming has been observed to some extent over regions spanning hundreds of millions of light-years, these streaming velocities are quite small compared to the speed of light, suggesting that the cosmological scale-expansion tracks closely across the universe. It is also possible that a feedback mechanism might exist that equalizes the cosmological expansion rate much like air pressure is being equalized by streaming air masses here on Earth.

A Whole New Ballgame

The link between GR and QT suggested by the SEC theory and the DIST process is both direct and clear. Hopefully, this connection will explain the following fundamental but previously poorly understood issues in QT:

- *The nature of the QT wave-functions:* Each particle is associated with a Compton oscillation in the metrics of spacetime, and this carrier wave is modulated by the QT wave-functions.

- *The particle/wave duality:* Modulation of the Compton oscillation during motion causes the de Broglie matter-wave. This clarifies the nature of the duality. Particle and wave are two sides of the same coin; particles might be standing waves in the metrics of spacetime that become modulated by motion.

- *Discrete quantum states:* A particle's energy determines its relativistic Compton frequency, which generates a corresponding QT resonance pattern. Therefore, there is a direct correspondence between energy states and distinct resonance states. Since particles prefer resonating states, the energy we observe is quantized. Note that these quantum states are a consequence of the oscillating nature of the particle.

- *Nonlocal action:* It might exist as determined by the GR geodesic. The non-local action does not occur in spacetime, but in the metrics of spacetime. There is no light-speed constraint for influence via the metrics.

- *Superluminal correlation:* Might also exist via the metrics of spacetime, which is a new "channel" of influence beyond spacetime.

Finally, I offer a comment on the effort to quantize the gravitational field. We have seen that the quantum world might result from the wavelike nature of particles and their Compton oscillation. Particles or fields that are not represented by such oscillation cannot be quantized. One example is the gravitational field expressing the curvature of spacetime due to matter-energy. If matter is formed as standing waves of the spacetime metrics, it is possible that matter-energy is generated by the nonlinear rather than by the linear part of the energy-momentum tensor of GR. This energy does not appear in the form of harmonic oscillation and therefore cannot be readily quantized. This might explain the difficulty encountered when trying to quantize gravity and could explain the much lower field strength of gravitational interaction.

I realize that this chapter on quantum theory does not cover all aspects of QT. However, I think that having a possible ontological explanation to our quantum world is better than no explanation at all. My hope is that it will inspire further investigation and lead to an improved appreciation of the world we inhabit.

Appendix IV firmly establishes the link between general relativity and quantum mechanics by deriving the deBroglie-Bohm pilot wave and the Schrödinger equation directly from general relativity.

CHAPTER 9

The SEC in Relation to Current Physics

SEVERAL YEARS HAVE NOW PASSED SINCE I FIRST REALIZED
HOW MANY WERE THE FALSE OPINIONS THAT MY YOUTH
TOOK TO BE TRUE, AND THUS HOW DOUBTFUL WERE ALL
THINGS I SUBSEQUENTLY BUILT UPON THESE OPINIONS.
—RENE DESCARTES

Hopefully you have now come to some understanding of the new physics I propose, but you might not yet have fully accepted it as your "new reality"; it might not yet have taken on an air of familiarity. Therefore, I think it would be helpful to briefly review some of the steps in the historic development of theoretical physics in relation to the new thinking presented in this book while highlighting where we might have gone astray in the past.

Some History

I have presented a new worldview that better agrees with astronomical observations and that explains and resolves several puzzling aspects of our world. This new worldview does not merely consist of a new cosmological model, but also implies a very different perspective on our existence. The beauty and internal consistency of the new cosmos model has hopefully become clear to the reader of this book. At first it might seem a bit difficult to assimilate, since the new physics introduces (and justifies) the existence of a dynamic spacetime scale and a new kind of process never before considered involving scale transition. The new thinking is therefore quite unfamiliar even to physicists.

However, the reader has seen how seemingly diverse subjects will coalesce by the help of the dynamic incremental scale transition (DIST) process and how it supports an overall worldview. This chapter will try to place the new physics based on a dynamic spacetime scale in the perspective of historic developments in theoretical physics and in relation to the currently accepted epistemology, which does not recognize the existence and importance of a dynamic spacetime scale. I highlight a few important issues, identifying obstacles that have hampered previous researchers in pursuing the line of reasoning I have followed. I show that most of these obstacles merely are preconceived ideas that we all have had about our world, ideas inherited from the past that unfortunately have been accepted as being true although they in fact may be wrong.

Not generally acknowledged, and sometimes plainly forgotten, is the fact our current knowledge is but a miniscule fraction of what humanity will learn in the future. Since we always build our scientific worldview on what we currently know, we must realize that it necessarily will be incomplete and sometimes even wrong. This is a self-evident human predicament, which is often overlooked. Unfortunately, there is no better alternative; we have to do the best we can basing it on what we know.

The same is true with the new worldview proposed in this book; the only thing we can say for sure at this time is that the SEC model with its DIST process appears to come closer to the truth than the SCM in that it better agrees with our instinctive impressions, with our observations, and with our experience.

I will try to point out where we might have made mistakes in the past. This review is important because it shows how a few seemingly innocuous assumptions made a long time ago may have hampered scientific development.

The Dawn of Western Science

The nature of motion has always been of fundamental interest to scientifically minded people. Aristotle believed that all free motion eventually has to come to rest. This remained the general belief until the time of Galileo, who made the revolutionary proposition that motion may continue unabated without external influences. This idea was later adopted by Newton and became his first law of motion.

Newton's first law of motion states that without any forces, a freely moving body will continue to move forever at the same direction and velocity. At the time of Newton, this was merely a conjecture that had not been proved, and I think Newton even realized that it might not be exactly true.

Newton's first law is closely related to his second law by which an applied force will accelerate a free, unrestrained, object. Without such a force, there would be no acceleration and no change in velocity in accordance with the first law. Newton's two laws of motion and his third law, which states that any force always encounters an equal and opposing force, are cornerstones at the foundation of classical as well as modern physics. However, there is a fourth, unnoticed, cornerstone that remains hidden, merely existing as an implicit preconception. It has been there from the beginning as part of the terrain upon which the edifice of science was built.

This previously unrecognized fundamental aspect of the world is the presumption that the 4D scale of spacetime is, and always has been, constant.

In the past, it never occurred to us that this might not be the case; we never thought that a changing scale of space and time could have revolutionary consequences. In fact, most of us have not even considered, or heard of, the possibility that the scale of objects might change dynamically, and that we all might participate in a process of cosmological scale-expansion.

Although at the time of Newton his first law was merely a conjecture, it has over time been elevated to becoming an uncontestable law of nature. Over centuries, a mighty structure of Science has been constructed with this law firmly at its foundation. The classical laws of conservation of energy and momentum are both based on Newton's first law.

However, we have seen that free motion might slowly diminish over billions of years due to cosmic drag. This violates Newton's first law. Such a tiny effect has in the past gone unnoticed, but recently some of its implications have been detected yet misinterpreted. One example is the decreasing pace of Universal Time, which generally is interpreted as being due to a slowing rotation of the Earth rather than (partly) being due to a slowly decreasing length of the year predicted by the SEC theory. You have already seen that evidence of this phenomenon recently has surfaced in the form of unexplainable observational drifts in the solar system, but this has largely been ignored. This is an example of where adherence to a generally accepted "law" may cause us to ignore unexpected observational discordances that could lead to new discoveries. This is a common human predicament; typically we discard, or explain away, discordant observations. Here is a quote:

> BY FAR THE MOST USUAL WAY OF HANDLING PHENOMENA SO NOVEL THAT THEY WOULD MAKE FOR A SERIOUS REARRANGEMENT OF OUR PRECONCEPTIONS IS TO IGNORE THEM ALTOGETHER, OR TO ABUSE THOSE WHO BEAR WITNESS FOR THEM.
>
> —WILLIAM JAMES

It is possible that an unproven hypothesis laid down some 400 years ago, which over time has become a generally accepted fact, profoundly has influenced the development of physics and our worldview in general, and has carried it in the wrong direction. Let's see how this might have come about.

A Crack in the Foundation

The first indication that something might be amiss with Newton's first law was the celebrated debate between Gottfried Wilhelm Leibniz and Samuel Clarke in which Clarke served as a spokesman for Newton. It is significant that Newton never directly confronted Leibniz in this debate; perhaps he was aware of the vulnerability of his conjectured first postulate that later became law.

The central theme of this debate concerned the existence of "absolute space," or in modern language, a cosmological reference frame (CRF). Here, we recognize an important aspect of science. Before a feature of the universe may be elevated to becoming law, its implications should be probed and challenged using observations as well as logical and philosophical reasoning. This is the essence of the scientific method, and the debate between Gottfried Leibniz and Samuel Clarke, speaking for Isaac Newton, is therefore of great epistemological as well as historical interest because it challenges the veracity of Newton's first and second laws. (Unfortunately, this important aspect of science seems to have been forgotten in the context of the now popular Big Bang model.)

Clarke (Newton) argued that an absolute reference frame must exist because of the existence of the inertial force. In his famous spinning bucket experiment, Newton observed that the surface of the water in the bucket becomes concave indicating that the water in the bucket "senses" rotation relative to something, and took this to mean that there must be some external reference for motion.

Leibniz, on the other hand, argued that according to Newton's laws of motion, all motion necessarily must be relative. By Newton's laws, all inertially moving coordinate frames are physically equivalent. However, Leibniz went one step further by arguing that inertial frames also should be conceptually and philosophically equivalent and that therefore that no CRF can exist, because there is no sufficient reason for it. Thus, if Newton's laws are absolutely correct and complete, logically no reference frame should exist. So, he concluded that either there is no CRF or Newton's laws are either wrong or incomplete. It is interesting to note that this question still remains open. It is commonly believed that both positions are true; Newton's laws are correct (with relativistic adjustments), and there is no cosmological reference frame. Until this day the crack in the foundation has remained open.

These two conflicting positions regarding the existence of absolute time and space expose a fundamental weakness of modern science; there is an irresolvable incongruity, which although it has not caused much trouble in the past now threatens the survival of the entire structure of theoretical physics.

Special relativity further exposes this inconsistency by its Twin Paradox. In spite of continued discussion and debate, the question of a cosmological reference frame has not been resolved over the 300 years since the time of Newton and Leibniz. However, from time to time, the question resurfaces. Einstein first sided with Leibniz, but later had to admit that empty space must have some inherent structure that explains the inertial force.

Another comment may be made regarding the Leibniz-Newton debate. Leibniz reasoned like the true mathematician he was when he deduced the logical consequences of Newton's laws, and challenged Newton to refute him. From Leibniz's point of view, Newton's spinning bucket observation indicated that Newton's laws were wrong or at least incomplete. However, the more pragmatic Newton thought that absolute space still might exist in spite of his laws.

We may compare this to the situation today. If general relativity (GR) were an absolutely correct and complete theory, we must conclude that the world was created in the Big Bang. But, a more pragmatic physicist, like Einstein himself was, did not accept the creation of the world from nothingness and concluded that there must be something we do not yet understand. This was also Einstein's position regarding quantum theory, but sadly he was ignored.

Einstein was strongly influenced by the German physicist Ernst Mach, who proposed that the stellar background serves as a CRF and that acceleration is to be measured in relation to this frame. In developing GR, Einstein had hoped that his new theory might resolve this issue, but found that this was not the case. Therefore, he reluctantly had to admit that his earlier abandonment of the aether might have been a mistake. This is what he said in his University of Leiden address in 1920 showing his ambivalence regarding the aether:

> *"But on the other hand there is a weighty argument to be adduced in favor of the aether hypothesis. To deny the aether is ultimately to assign that empty space has no physical qualities whatever. The fundamental facts*

of mechanics do not harmonise with this view. For the mechanical behavior of a corporeal system hovering freely in empty space depends not only on relative positions (distances) and relative velocities, but also on its state of rotation, which physically may be taken as a characteristic not appertaining to the system in itself."

At the end of his life, Einstein was convinced that spacetime was a new form of aether that somehow served as a reference for Inertia. However, he still believed that all inertial frames were equivalent. He was right on both accounts, but it now appears that an important aspect was missing: the dynamic metrical scale.

If, at the outset, Newton instead had taken into account the possibility that all relative motion might slow down with time (which could be very long), the debate with Leibniz would never have taken place, because it would have implied that moving matter eventually would tend to a state of relative rest, which implicitly would define absolute space as the frame toward which all motion converges. But Newton did not do this, probably because he could not justify making the assumption that motion dissipates.

By the end of the nineteenth century, the physics community was more or less aligned with Newton in believing that the stationary aether representing absolute space must exist and be the carrier of light. However, after Einstein's SR paper appeared in 1905, the aether idea was gradually abandoned in favor of Leibniz's relativistic point of view.

It now appears that this might have been a mistake with very serious consequences.

Enter Special Relativity

As you already know, Einstein based his SR theory on two postulates, which I restate here for convenience:

- All inertial frames are physically equivalent in the sense that the laws of physics hold equally true in them.

- The speed of light is the same in these frames.

However, in his 1905 paper, Einstein implicitly also made an additional assumption: he assumed that the coordinates of a moving reference frame may be expressed mathematically as functions of the coordinates of a stationary reference frame by the Lorentz transformation, which at the time already had been proposed by two dominant personalities in science, Henrik Lorentz and Henri Poincaré. Therefore it is not surprising that Einstein, who was merely a patent clerk at the time, adopted the Lorentz transformation.

However, he also assumed that coordinate increments in inertial frames as given by the LT have the same meaning, and therefore that coordinate increments in a moving frame may be directly compared to those of a stationary frame. Thus, he believed that the coordinate increments of a moving frame, as seen from a stationary frame, are directly comparable to those experienced by an observer, who is stationary in the moving frame. This is not necessarily true, because although inertial observers all may experience their own local frames the same way, it is possible that they do not experience moving frames the same way as their local frames. I don't think the possibility that coordinate increments might have different *relative* meanings ever occurred to Einstein (or anyone else). However, it is possible that a moving frame will appear to have different metrics than what is locally experienced in a stationary frame.

In other words, the relativity of inertial frames may extend to their metrical scales.

I discovered this hidden property of the universe when searching for the origin of the inertial force, which may be explained as a curved spacetime phenomenon if the scale of spacetime were to change during acceleration. Therefore, a dynamic 4D spacetime scale could be a previously hidden aspect of all motion. And, when it turned out that the same inertial scale factor that explains the inertial force also appears in SR, it suggested a connection between inertia and SR, which provided further support for the proposition that the metrical scale might be relative.

This would mean that the equivalence of coordinate increments as expressed by the LT is an unjustified, erroneous assumption, which has caused a lot of puzzling problems, most prominently the infamous Twin Paradox whereby twins who part on different journeys upon reconvening both conclude that their sibling should be younger.

I doubt that an independently thinking reader familiar with SR really is comfortable with the Twin Paradox, in particular if this reader agrees that acceleration should have no effect on elapsed time intervals. Obviously, the twins' clocks must always run at the same pace, since they both are at rest in their local frames of reference, which by SR are physically equivalent. This is the essence of relativity. For symmetry reasons, this leads to the obvious conclusion that the clocks must agree when compared after travel, regardless of which twin did the traveling (or if both travel), which disagrees with SR. Yet, it appears that some people still believes that the pace of time actually changes during motion.

The Twin Paradox may be resolved if we accept that *coordinates related by the LT might not have the same meaning;* their metrics might differ. The implication of this would be that although important properties of moving frames are captured by the LT, the coordinates of a moving frame given by the LT are not necessarily identical to those experienced locally by a co-moving observer. In other words, a stationary observer experiences the coordinates in a moving frame differently from those seen by the moving observer at rest in her local moving frame.

To explain how this situation might arise, we need some additional background, which was not available to Einstein in 1905.

Enter General Relativity

GR was developed by Einstein and published in its final form in a famous paper [Einstein 1915], ten years after his paper on SR. Had the order of these two papers been reversed, the development of physics might have taken a different and more favorable path. I will not go into the details of GR, which are mathematically formidable, but will only review a few salient aspects of the theory needed for this discussion.

The most important aspect of GR is its description of the world in terms of 4D geometry: one temporal and three spatial coordinates. This 4D representation was proposed by the German mathematician Hermann Minkowski, one of Einstein's teachers, who found that the Lorentz transformation preserves distances in a 4D space and that it may be seen as a rotation in this space.

On September 21, 1908, Hermann Minkowski began his talk at the 80th Assembly of German Natural Scientists and Physicians:

"The views of space and time which I wish to lay before you have sprung from the soil of experimental physics, and therein lies their strength. They are radical. Henceforth space by itself, and time by itself, are doomed to fade away into mere shadows, and only a kind of union of the two will preserve an independent reality."

This conclusion seemed justified at the time since the LT's temporal coordinate relation mixes time with space, suggesting that time and space cannot be separated. Therefore, it seemed that it was no longer possible to treat time and space as being different and separate entities. (However, as we saw, this is not correct, because of the existence of a cosmological reference frame. And, the cosmological scale-expansion defines an absolute time in the form of atomic time.)

With this development, space and time were merged into 4D spacetime. GR makes use of this formulation and expresses physical properties of the world via differential geometry defined by a line-element. It describes the relationship between coordinate increments in space and time in terms of metrical components for the coordinates.

It is important to note that GR provides a *static 4D representation of the world*, which does not distinguish between the past and the future. Therefore, it cannot model, or explain, the progression of time, or describe motion *as a physical process*. Like a map describes a geographical area but not the process of traveling in this area, GR maps the world but not our travel in this world via the progression of time. In retrospect, this should tell us that GR cannot be the last word.

GR models a gravitational field by metrics that depend on location. Thus, the meaning of relative increments in space and time may change with the location in 4D spacetime. Einstein introduced the new idea that the "density of spacetime", as expressed by the metrics of space and time, may be used to model gravitation and most laws of physics. This was a truly remarkable idea, although other scientists before him had also considered modeling the world by geometry—for example, the German mathematician Bernhard Riemann, who considered non-Euclidian three-dimensional differential geometry. However, Einstein was first in exploring 4D pseudo-Euclidian geometry in his GR theory.

Thus, geometric properties of spacetime can model the gravitational force and explain its mysterious action at a distance as being caused by spacetime curvature induced by changing metrics. Like the slope of a hill causes a downhill pulling force, the gravitational field is thought of as being due to changing the metrics of spacetime, which causes the gravitational force. Einstein called GR a theory of gravitation, but it now appears that GR is a very important step toward a much improved description of the world covering other aspects than gravitation.

In this context, GR may be seen as the culmination of a long line of development based on mathematical differential methods that originated with Newton and Leibniz. By this differential approach, the world is modeled as a continuous manifold in which increments in space and time may become arbitrary small. GR might be the last major contribution along this line of development. It bridges the past and the future by recognizing the important but previously unsuspected role played by the metrics of spacetime, but GR still retains continuity. Nowadays GR together with QT are the two dominant theories in physics. Ironically, the domain of QT may be the metrics of spacetime. We might say that GR and QT are like estranged siblings who grew up in different foster families, but now will find that they are closely related.

This brief historic review sets the stage for the new development presented in this book.

Spacetime Scale-Equivalence

There is a fundamental symmetry of the universe that in the past has not been properly acknowledged in spite of its importance. Physics, at least as I once learned it, ignores this symmetry; but it could be the key to explaining the cosmos, including its quantum aspects, the phenomenon of inertia, and the progression of time. This hidden property of the universe is:

There is no preferred 4D scale for existence; the cosmological spacetime scale may change.

In GR, a different scale may be modeled by multiplying all terms in the line-element by the same constant scale-factor, and we find that this will not change Einstein's field equations at all. This means that scaled spacetimes are physically equivalent; they are *scale-equivalent*.

Thus, the 4D scale is a free parameter.

Although things obviously have a scale, this cosmological scale is inaccessible by local measurements and would not be of interest to us unless it changes with location in space or time. However, we have seen that this actually could be the case and that it would have revolutionary consequences for physics.

The fourth cornerstone of current science is the unjustified presumption that the scale of space and time always is, and always has been, fixed.

If the scale is not fixed but changes incrementally, this new type of process will not be "seen" by GR, since the field equations with their physical description of the universe do not change. This implies that GR cannot model processes involving a dynamic 4D scale as *experienced by an observer participating in the process*, which might explain why this new type of scale transition process has not been noticed or explored in the past.

SR in view of GR and Inertia

This new insight also sheds new light on SR. The LT, which modifies the classical Galilean transformation, attempts to express the position in space and time of a moving object in relation to a fixed coordinate system, subject to additional constraints posed by Einstein's two SR postulates. These two postulates mean that the Galilean transformation had to be revised to include a new term in the temporal transformation between frames, which depend on location. The LT is believed to give the correct model of two coordinate frames in relative motion.

However, if the LT instead is viewed as being a coordinate transformation in GR, it will no longer model motion but instead relate 4D geometries as experienced by observers in relative motion. From this geometric point of view, the LT will conserve all physics simply because it conserves the line-element in GR. This interpretation of SR in the context of GR will automatically satisfy the two postulates of SR, a theory that was developed before GR.

Hence, the LT modifies the Galilean transformation to satisfy the two SR postulates, and since it preserves Einstein's GR equations, it also conserves all physics including Maxwell's equations. In fact, in retrospect it seems likely that the reason why the LT transformation works so well is that it preserves Einstein's field equations.

However, there is something strange with this geometric interpretation because according to the LT, all inertial frames will have the same geometry, which raises the question of what might be causing the inertial force. If inertia is a phenomenon akin to gravitation, as assumed by Einstein, the inertial force should, like the gravitational force, arise from spacetime curvature caused by changing spacetime metrics. One wonders how this may be possible if all inertial frames have the same identical spacetime geometry. Therefore, as an unfortunate and unintended consequence of SR, the explanation to inertia has been missing.

A small, seemingly innocuous crack in the foundation implied by Newton's first law has reached out from the past via SR and is presently obstructing further development.

The Downside of Relativity

The idea that inertial frames are conceptually on equal footing, as initially advocated by Leibniz, has had deep philosophical consequences, because it collides with observable facts. For example, relativity of moving frames together with Newton's first law implies that there should be no cosmological reference frame, yet our observations tell us that galaxies are more or less stationary with relative velocities less than a fraction of a percent of the speed of light. How may this be explained? Simulations show that these velocities ought to be a lot higher if the SCM were true. Also, the phenomenon of inertia tells us that particles somehow "know" that they are accelerating and we might rightfully ask:"Relative to what?" Furthermore, QT demands a temporal reference background, which must exist to allow nonlocal influences faster than the speed of light. Therefore it appears that Newton's "absolute space" should exist.

However, nowadays most scientists believe that a good scientific theory should be mathematically formulated independent of the choice of coordinates, since this is a feature of GR. The idea here is that the underlying physics does not depend on the coordinates we choose to describe it and that therefore, from a purely mathematical point of view, the choice of coordinates is arbitrary. One example is the use of rectangular coordinates or polar coordinates; they may both be used to define the location of a point in space. Therefore by GR, the choice of coordinates is arbitrary as long as they may be related by a continuous variable transformation. This property of GR is called *covariance*.

However, in practical applications we are using specific coordinates. This is the case in quantum mechanics and in cosmological theories, which both imply the existence of a common temporal reference. Also, by the present adoption of atomic time we have in effect admitted that the choice of the temporal coordinate is not arbitrary. This has led to a rather confusing situation.

The fact that GR does not provide a preferred choice of coordinates has complicated the task of modeling the world by mathematics, because our description of the universe strongly depends on the coordinates we choose. For example, according to GR, two physically equivalent cosmological line-elements may co-exist, with one of them implying that there is a beginning of cosmological time while the other implies eternal existence. GR does not tell us which one of these line-elements is right since their physics is the same, which indicates that something surely must be wrong; they cannot both be right. It appears that GR is *too* general. Perhaps things would have been easier if we had accepted that a cosmological reference frame exists, although it might remain hidden in experiments, like the one of Michelson-Morley, that measure directional differences in the speed of light.

With the discovery of atomic time, Nature has revealed its own temporal preference, which clearly shows that the choice of coordinates is not arbitrary. In fact, the in chapter 5 mentioned observational discrepancies in the solar system show that optical observations based on atomic time differ from planetary ephemerides using the fitted time-base T_{eph} that is computed based on Newton's laws of motion and his law of gravitation with relativistic refinements. Obviously the choice of coordinates *does* matter.

Therefore, it appears that the purely relativistic worldview we adopted during the past century has not helped clarify or simplify our understanding. Rather, it may have imposed unrealistic constraints that are obstructing progress.

General Relativity Cannot Model the Cosmos

In retrospect we may find that the adoption of a cosmological model based on the Big Bang creation idea was one of the most unfortunate developments in the twentieth century. Cosmology has always been at the core of scientific endeavor and has historically motivated much of its development. The reason is clear: the ultimate lofty goal of science is to improve our understanding of the world. Therefore, describing the universe is of great interest to most of us. The mistake we made with the Big Bang idea could therefore be characterized as a tragedy. In the future, it will likely be compared to the flat Earth model and the geocentric Ptolemaic system.

In retrospect, it is perhaps not surprising that a cosmos model sanctioning the creation of the world gained acceptance in the West, because it agrees with the deeply held religious belief. In fact, one of the main proponents of the Big Bang model was George le Maître, a Catholic priest. However, from a purely scientific point of view, it is troubling that the universe currently is being modeled by GR, a theory that cannot explain the progression of time, which arguably is the most important aspect of all existence. It seems obvious that a theory that cannot explain the progression of time clearly must fall short in attempting to model the universe.

The ancient Greek philosopher Parmenides may have been right in asserting that something can never be created from nothingness and, as a consequence, that existence must be eternal. And, if existence is eternal, the universe must always have remained physically the same, and the spacetime geometry cannot have changed with time. In standard physics, this would imply a static universe where nothing ever changes, which does not agree with our experience. However, this is what GR seems to tell us, which has led some people in science to argue that the progression of time merely is an illusion; it is something strange beyond known physics.

When confronting something unexplainable, this attitude is not unusual among theoretical physicists as well as among the public in general. There is a tendency to cling to established beliefs rather than relying on observations and independent common sense reasoning. Someone who has spent a lifetime immersed in a certain worldview and who has taught it to others might naturally find it very hard to accept that it all might have been a mistake. It is much easier to hold on to an outdated belief even when the evidence against it is overwhelming, as judged by an unbiased observer.

Einstein spent his last 30 years trying to make further headway along the path of differential geometry and continuous manifolds. However, in a letter he wrote to his friend Michele Bosso about a year before his death, he expressed reservations about the very edifice he helped to create. This is what he said:

> *I consider it quite possible that physics cannot be based on the field concept, that is, on continuous structures. Then nothing remains of my entire castle in the air, including the theory of gravitation, but also nothing of the rest of modern physics.*

In this, Einstein was perhaps a bit too pessimistic; without him, we would not have recognized the important role played by spacetime geometry. He took this idea as far as possible without abandoning his conviction that the world is basically four-dimensional and continuous. However, this doomed his later efforts.

A Dynamic Scale Process to the Rescue

There are several troubling aspects with modern physics: astronomical observations do not agree with the SCM's predictions; there are problems with SR as evidenced by the Twin Paradox; the absence of a cosmological temporal reference frame is mysterious; explanations to the progression of time and to the origin of inertia are missing; and QT is incompatible with GR. However, it is possible that all these seemingly disparate problem areas might have the same root cause; it is likely that these difficulties and incongruities might be resolved by considering the possibility that the scale of spacetime might change dynamically. A dynamic scale would introduce a new "player" in the game of physics that might help us to better understand the universe.

We should note that *4D cosmological scale-expansion cannot be modeled by GR*, since GR does not provide a way of describing scale-expansion *as experienced by an observer who participates in this expansion, like we all do*. Relative to such an observer, the 4D spacetime geometry will always remain the same. GR cannot model this situation because GR represents a static 4D geometry with a constant reference interval. But if we generalize GR to admit discrete scale adjustments, the SEC may be modeled by GR, since Einstein's field equations remain the same with discrete scale adjustments. This suggests the existence of a new, semi-continuous physical process involving dynamic incremental scale transition (DIST).

Here it is important to recognize that the reason we ended up with the unfortunate SCM theory is that GR cannot model scale-expansion. Therefore, already known and accepted science has prevented us from finding a new and better model. This is a consequence of the scientific method; it forces us to try to explain the world by what currently is considered known, even if what we know is inadequate and unable to explain it. Other kinds of explanations are not acceptable.

Thus, the SCM is the best we can do based on GR.

In the context of SR, we already saw that acceleration might change the scale of spacetime via the DIST process and that this could explain inertia. With this new kind of process, spacetime would always remain the same relative to an observer participating in the acceleration, but the scale of other moving inertial frames would appear contracted in relation to this observer. This is a symmetric situation; all inertial observers would see smaller scales in other moving inertial frames. Also, an observer in the SEC sees smaller scales in earlier epochs. This somewhat unfamiliar situation may be explained by a dynamic scale and by a process that takes place beyond the four dimensions of spacetime.

Of course, this process is something new and unfamiliar, and you might perhaps wonder if it really is true. This question cannot be answered conclusively; instead, we should ask if it would help explain the world. If this is the case, it should be tentatively accepted and hopefully be verified by further investigation.

It would be a mistake to reject the DIST process simply because it is unfamiliar.

The DIST Process as a New Cornerstone

By considering the possibility that the universe makes use of scale-equivalence, which preserves all physics, we find a new cosmos model that resolves many cosmological mysteries and explains puzzling observations. For example, the incrementally expanding scale makes time progress, explaining the enigmatic progression of time. This problem was early recognized by Parmenides and his follower Zeno, who wondered how things can ever change in eternal existence. In the scale-expanding cosmos, the 4D spacetime geometry does not change as perceived by a co-expanding observer, but the scale does. Although spacetime may be modeled by a continuous GR manifold, the scale adjustment could be a discrete process modeled by dynamic incremental scale transition.

From a purely philosophical point of view, we might wonder why Nature would make use of a dynamic scale-factor. One answer could be that scale-equivalence is the most fundamental symmetry of all since it conserves *everything* four-dimensionally. It does not only conserve our spatial world but also dynamic processes involving flows of energy. In other words, it conserves a cosmos in perpetual motion with the expanding temporal scale being the engine of all existence.

A second answer could be that all particles are resonating metrical oscillations in spacetime that depend on a uniform, isotropic speed of light for their existence. Therefore, particles may preserve the conditions necessary for their existence by locally adjusting their spacetime metrics. This does not mean that a particle somehow changes the "underlying cosmological fabric" of spacetime; spacetime may remain the same relative to observers in other frames. Instead it is achieved by locally *redefining* distance and temporal increments. Every particle (observer) may then form its own perspective of the world. Consequently, an observer in motion may find the speed of light to be constant in relation to her local reference frame and all laws of physics to hold true as measured by local measuring rods and clocks. Yet, this observer's local spacetime geometry might differ relative to those observed in other inertial frames. In the vocabulary of GR, they would be in different four-dimensional "manifolds" embedded in a higher dimensional space; inertial frames would *not* be covariant. This possibility would extend "relativity" to 4D spacetime geometries, each inertial frame giving its own interpretation of the world.

Visualizing the Cosmological Scale-expansion

When we think of "expansion," we visualize something that grows with time. In a sense, this is also true for the SEC, but this is the perspective of a *fictional* observer, for whom the scale does not expand, who is located "outside the 4D universe" in another dimension. However, the perspective of a co-expanding observer is quite different, because such an observer will locally not experience any expansion with time but rather cyclic or vibratory processes in the spacetime metrics. Geometrically, on the macro scale, space will always appear locally the same at each instant, but on a submicroscopic level the scale-expansion announces its presence as oscillation in the metrics, explaining the quantum world.

Although the "outer-worldly" view of the scale-expansion might help us understand this new expansion mode, it is not the perspective we actually have. Before we can really appreciate the beauty of our dynamic, eternally scale-expanding world, we must learn to visualize it "from the inside", realizing that we all are participating in a perpetual flow of existence.

The Connection with Quantum Theory

Another implication of semi-incremental scale-expansion is that it could induce oscillations in the metrics of spacetime, which could explain our quantum world. *QT may be derived from GR if the metrical scale oscillates.*

It is well-known that particles are associated with waves at the so-called Compton frequency and that this frequency is proportional the mass-energy of the particles. However, the nature and origin of these waves have remained unknown.

If this Compton oscillation were in the scale of spacetime, it would provide the missing connection between GR and QT.

If this were the case, the de Broglie matter-wave would become a natural consequence of motion. Furthermore, the de Broglie–Bohm pilot function that seems to guide quantum particles may then be derived from the geodesic equation in GR, and the Schrödinger equation may also be derived from GR by setting its Ricci scalar equal to zero. This would together with some random disturbance explain the quantum world. The scale oscillation that sustains all particles could be energized by the incremental cosmological scale-expansion, which also would generate oscillating random disturbance at extremely high frequencies.

This provides a firm link between cosmology and quantum theory via the DIST process and GR, a connection that previously has been missing [Masreliez, 2005a].

Summarizing this Chapter

The theories and assumptions that over the past four hundred years have formed the foundation of modern physics have obviously been quite remarkable. However, in retrospect it now appears that they might have been *too* persuasive; their general acceptance has become dogmatic, which might have prevented the discovery of a new fundamental feature of the universe - its dynamic scale.

When Isaac Newton laid down his first postulate, later to become law, according to which constant motion will continue unabated forever in the absence of external forces, it was one of three postulates upon which he constructed a beautiful and very successful theory of motion. The success of Newton's approach combined with his brilliant presentation led to a widespread general acceptance of his theory. The first law became the conservation of momentum, which soon was treated as being absolutely true; it became a *law of nature*. Likewise, the assumption that the laws of physics are the same independent of time became the conservation of energy and the first law of thermodynamics. However, these two conservation laws have never been formally proven.

We are now beginning to see unexplainable discrepancies in the motion of the planets (planetary acceleration) and in signals received from space probes (Pioneer Anomaly), which indicate that Newton's laws might not be entirely correct. This is, of course, difficult to accept because it would obsolete celestial mechanics.

Based on Newton's three laws of motion, a mighty theoretical structure laying down different aspects of motion was constructed during the eighteenth and nineteenth centuries by several mathematical physicists. Names like Pierre Simon Laplace (1749–1827), Joseph Louis Lagrange (1736–1813), and Adrian Marie Legendre (1752–1833) come to mind. Another well-known contemporary was the Irishman Sir William Rowan Hamilton (1805–1865). Much of our understanding and treatment of motion was developed by such pioneers; for example, the minimization of the so-called Lagrangian is now a standard approach, which often is the beginning of many theoretical investigations without remembering its ancient origin in Newton's laws of motion. Consequently, if something is amiss with Newton's laws, we might perhaps question the validity of some of the results of modern physics.

A major event in theoretical physics was the publication of Einstein's SR theory, which initially was met by considerable skepticism because it introduced something new and different by mixing spatial and temporal coordinates. SR is based on the older Galilean transformation, which independently relates the spatial and temporal coordinates. Before SR entered onto the scene, all visualization of motion assumed that when a moving object changes location in space, each consecutive position corresponds to a certain time instant of an absolute time that exists and is the same for all observers whether or not in motion. Thus, absolute time was an independent parameter that was the same everywhere at any instant. (It is ironic that we now may return to the same understanding.)

This is no longer true in SR, a fact that caused much consternation. The widespread skepticism of SR was further deepened when it was discovered that the theory implies logical contradictions such as the Twin Paradox. The debate on SR continues to this day on the Internet, for example by members of the National Philosophic Alliance (NPA) and in the journal Galilean Electrodynamics.

The main strength of SR is that it is based on two simple postulates, which together with the assumption that all inertial coordinate frames are physically equivalent, led Einstein to the Lorentz transformation. But there is a hidden problem with his derivation; there is an aspect of motion that he did not, and could not, have envisioned, and which has not been properly recognized until now. Motion changes not only the location of an object in space and time but also its metrical scale of spacetime in a relative sense. The dynamic scale provides a new degree of freedom, which allows validity of the two SR postulates and the equivalence between inertial frames while resolving the Twin Paradox. However, it leads to a different theory that also explains the phenomenon of Inertia.

So, we may conclude that although certain aspects of SR are correct, this theory is conceptually misleading because it does not recognize the role played by dynamic spacetime metrics. With this unfortunate omission, the acceptance of SR with its Lorentz transformation has in effect become an obstacle in the further development of physics.

The next major advancement in theoretical physics came with Einstein's GR theory, which modeled the world using 4D differential geometry. GR describes the geometrical properties at various locations in 4D spacetime by comparing the local metrics at every location with a reference interval valid across the entire space. We might say that this reference interval (commonly denoted *ds*) provides a common reference for spacetime in the sense that all increments are compared to *ds*. It equals "proper time," which is time for an object at inertial rest without gravitational fields.

Within a GR manifold, the geometry is relative, and Einstein's field equations do not "notice" its scale. The scale-factor appears in both numerator and denominator in the so-called "Christoffel symbol" and therefore cancels out.

There are two problems with this. First, processes might exist that alter the scale dynamically, which cannot be modeled by GR. Second, by GR time is merely one of four static geometric coordinates, which means that GR cannot model the *progression* of time or the *process* of motion. *Therefore, GR cannot model the universe.*

Note that all these shortcomings are caused by not recognizing that there is a missing "dimension" that is reflected in the progression of time, which might participate in all kinds of motion.

The review in this chapter is briefly summarized in the following table:

Scientific theories	Interpretations	Consequences
Newton's first law	Conservation of momentum (spatial symmetry) and energy (temporal symmetry).	No cosmological reference frame. First law of thermodynamics. Ever increasing cosmological entropy.

Special relativity	The coordinates given by the LT are assumed to have the same relative metrics.	The Twin Paradox. Impossible to explain Inertia as a curved spacetime phenomenon.
General relativity	Assumes that the world may be modeled by 4D differential geometry.	There is no explanation to the progression of time. Leads to the Big Bang model with its creation event.
New theory based on dynamic spacetime metrics	Motion in time or space involves dynamic spacetime metrics. The expanding scale models the progression of time. The Zero Point Field and the quantum mechanical wave-functions are oscillation of the spacetime metrics. Inertia is caused by dynamic metrics.	Cosmic drag induces a cosmological reference frame. An expanding scale agrees with all observations and explains the universe. Eliminates cosmological creation and allows eternal existence. Explains the origin of Inertia. The Twin Paradox is resolved.

CHAPTER 10

Bits and Pieces

FEW ARE THOSE WHO SEE WITH THEIR OWN EYES AND FEEL WITH THEIR OWN HEARTS.

—ALBERT EINSTEIN

This book came together piece by piece over many years, during which I from time to time jotted down my thoughts as they occurred to me. Some of these scattered comments are presented in this chapter. There is no particular order to these ruminations and the reader should take them as separate, unrelated samples of my thinking over the years. Hopefully they will help complete the picture.

Aristotle, Motion, and Inertia

People in the ancient Greece wondered why a projectile—for example, an arrow shot from a bow, would continue to move after leaving the string. Since there no longer is any force pushing the arrow, it seems that it immediately ought to fall to the ground. Aristotle argued that the arrow continues flying because the string of the bow puts in motion not only the arrow but also the air surrounding it. Thus, he believed that this co-moving air somehow carries the arrow with it, with the arrow at rest relative to it. He came remarkably close to the right explanation!

However, this explanation was successfully refuted by Galileo, who thought that something inherent to the arrow rather than its surrounding air must cause it to continue on its trajectory. He noted that a cannonball dropped from the top of a mast on a ship will always hit the deck below at the same spot regardless of whether or not the ship is moving, provided the motion is uniform. He further generalized this to include objects dropped anywhere on Earth; they always fall straight down regardless of the fact that the Earth rotates and moves around the Sun. With this reasoning, he refuted the ancient argument that vertical falls prove that the Earth does not move.

Of course, nowadays we explain this situation by the presence of inertia and the law of conservation of momentum. But this does not really explain what is going on either. Inertia and momentum are concepts we have *invented*, which serve to help explain why the arrow continues flying. They are based on Newton's first law, which Newton postulated but never explained. It says that a body without any applied force will continue in uniform motion. Although we nowadays think of this law as being a scientific explanation and that no further explanation is needed, the invention of new concepts like inertia to explain things we don't understand do not really offer any explanation at all; we could equally well have argued that an invisible fairy named Sally carries the arrow in her left hand!

Of course, the concepts of inertia and momentum serve to explain a great number of other phenomena and have therefore gained general acceptance. They are part of scientific epistemology. But what inertia really is and why it should exist at all has eluded us; nowadays, sadly few give this question any further thought. This should cause us to pause and think; we have built our Western science on an unproven concept of unknown origin and have conveniently forgotten that Newton's second law $F = am$, which says that a force is needed to accelerate an object, merely is a postulate based on a conjecture.

Aristotle, it seems, could have been right in that something moves with the arrow and keeps it flying. I think that in this respect Aristotle actually *was* right! However, it is not the air that moves with the arrow, but its local spacetime geometry. In order to understand this we have to take a new look at what constitutes motion. In the chapter on inertia, I set the stage by showing that motion is not a simple concept; in fact, it is impossible to explain it as a continuous process in fixed spacetime. Rather, I suggested that all moving objects carry with them their own local version of spacetime, possibly because they need to conserve the local geometry in order to sustain their existence.

Acceleration generated by the bow's string may change the local spacetime geometry and allow the arrow to continue on its flight. Acceleration must therefore induce spacetime curvature, which explains the inertial force as being akin to the gravitational force; it is a curved spacetime phenomenon. And, after accelerating and adjusting its local spacetime to the new velocity, the arrow will continue in its flight, being at rest in its local co-moving spacetime. Thus, we find that Aristotle would have been right with the co-moving air replaced by co-moving spacetime geometry.

The main thing to realize here is that matter might not exist *in* some kind of aether, or in a fixed background spacetime as is the traditional thinking, but that a particle may adjust its local spacetime geometry *"at will"* depending on its motion. This implies the co-existence of innumerable 4D spacetime geometries existing in a cosmos of more than four dimensions.

There is another point to be made here. If spacetime changes with acceleration, a particle must somehow sense that the properties of spacetime change. It would be hard to understand how this can be possible without the existence of some kind of background reference. Fortunately, the SEC theory provides just such a reference that is present locally and permeates everywhere at all levels across the cosmos; it is the expanding spacetime scale. The cosmological reference is no longer defined by distant matter like by Mach's principle, but it is always present, right now, here and everywhere, being induced by the dynamically expanding scale.

The cosmological scale-expansion may be directly reflected in atomic time. Inertia may be a direct consequence of preserving the pace of atomic time when the velocity changes. If this is the case Einstein's two SR postulates will together with atomic time require inertia. This makes sense; we saw that SR's loss of absolute time makes it impossible to explain inertia.

A Comment on Modeling Gravitation

By introducing the dynamic scale and the DIST process, we can model acceleration by a sequence of discrete 4D Minkowskian manifolds. At each location on an accelerating trajectory, the dynamic inertial scale factor models the inertial force by the geodesic of GR. This allows the accelerating particle to retain its flat Minkowskian spacetime locally at each instant by adjusting the dynamic scale factor.

If Inertia and Gravitation are both curved spacetime phenomena, this approach should also apply to Gravitation; it should be possible to model it by a sequence of Minkowskian spacetime manifolds. The temporal scale given by Schwarzschild's line-element for a spherically symmetric gravitational field is $(1 - 2GM/rc^2)$ and since the corresponding inertial metric is the inertial scale factor $1 - (v/c)^2$ we find the correspondence

$$\frac{v^2}{2} = \frac{GM}{r}$$

Here the velocity is the same as that of a freely falling particle, and expresses a relationship between the gravitational potential and the inertial potential at each instant. Comparing this situation with our model of inertia, we find that the gravitational field may be thought of as being caused by accelerating spacetime "streaming into" the mass that creates the gravitational field. Thus, it appears that the cosmological scale-expansion somehow induces this accelerating, streaming spacetime.

This image suggests a new aspect of motion via an incrementally changing 4D scale like was done both in modeling the cosmological expansion and when modeling inertia. With this in mind we may think of a particle suspended in a gravitational field—for example, here on the Earth's surface, as being accelerated relative to the locally streaming spacetime, thus being in a situation similar to an accelerating particle in empty space.

The fact that all particles of matter seem to consist of subassemblies of even smaller energy packets in decreasing hierarchies suggests that these subatomic entities might all exist in locally scaled Minkowskian spacetimes that dynamically adjust their relative scales in an incessant energy flow.

By this new way of visualizing acceleration and gravitation the new scale dimension takes on an active role in all motion. The cosmologically changing scale corresponds to the progression of time, and we realize that the old way of treating motion as a world-line in 4D spacetime is inadequate because it does not model the *process* of motion.

The dynamic scale could offer a new aspect of existence that improves our description and deepens our understanding of the universe.

An Implication of Cosmic Drag

I would like to briefly mention a few additional interesting consequences of cosmic drag.

Since the planets spiral toward the Sun, the Earth was once farther away from the Sun, and its slow and steady approach over billions of years has gradually warmed its surface and made life on it possible. Five hundred million years ago, the Earth was about 5 percent farther away from the Sun. If the Sun was as luminous as it is today, the radiation received by the Earth was about 10 percent less than it is today, which would have had strong impact on life on Earth at the time. In the future, we will be heading for unavoidable warming that goes far beyond the current global warming. We might therefore be living in a window of time with a temperature that favors us human beings. Although this might limit the duration of existence for different life-forms on the Earth including us humans, it will greatly increase the likelihood for life elsewhere in the universe. Since the planets in all solar systems slowly drift inwardly, life becomes possible on many more planets in numerous solar systems. At some time they will be at a distance from their mother suns where the temperature is just right for biological life.

A Few Comments on Simultaneity

Let me add a few comments on the concept of simultaneity. Currently the belief is widespread that simultaneity of events depends on the observer's reference frame; according to SR, events that are simultaneous in one inertial frame may not be simultaneous in another frame. Although this belief is by now deeply ingrained in both physics and philosophy, it could be an unfortunate misconception, which has caused much unnecessary concern and debate over the past hundred years.

It may simply not be true that events that are simultaneous in one inertial frame are not simultaneous in other inertial frames. The reason is simple: *atomic clocks always run at the same pace* in inertial frames. This means that we may directly compare clock readings and define events as being simultaneous if their clock readings are the same. This new definition of simultaneity, which coincides with the classical definition, becomes possible by the cosmological scale-expansion, which plays the role of a master clock for the universe, as reflected by the progression of atomic time.

Of course, this does not agree with SR, which bases its definition of simultaneity on the assumption that the velocity of light is constant and the same on all inertial frames but does not take into account the dynamic scale or the cosmological expansion. We saw that as a consequence the temporal coordinates related by the Lorentz transformation have different metrics, which makes a direct comparison in the context of SR impossible. Even if the *observed* readings of a moving clock disagree with a local clock, it does not mean that the moving clock actually runs at a different pace. What we observe could be a distorted view; it might merely be a projection of the moving frame onto the local frame.

We may avoid the inconsistencies implied by SR by embracing a new aspect of existence. Although the argument may be made that the introduction of a dynamic relative scale complicates the description of the universe (which is not true), it eliminates inconsistencies and contradictions. Obviously, a theory with contradictions will never be acceptable. And papering over its flaws will never make it right. However, if a new theory is found that eliminates problems, it should be preferred, even if it initially appears more complicated.

I will further justify my contention that absolute time exists by assuming that all particles are standing waves. How a particle actually is formed and what constitutes these standing waves is immaterial to the following line of reasoning.

When a particle moves, it might preserve its oscillation properties *including the oscillation frequency* as illustrated in figure 21, panels (a) and (b), where IRF refers to Inertial Reference Frame.

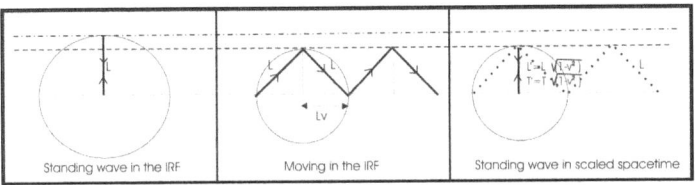

Figure 21: Standing wave particle in motion

We see that preserving a constant (atomic) pace of time would imply that the scale perpendicular to the motion contracts, since the path length of the light remains unchanged in panel (b). This corresponds to a reduced scale in directions perpendicular to the motion and explains how *the inertial scale factor expresses a particle's response when changing the velocity*. If the pace of time is also adjusted by the same factor, all four coordinates are scaled equally as in the inertial line-element. This is illustrated in panel (c). By scale-equivalence, all physical properties will then be preserved.

Assume that after a brief acceleration-boost, the particle has obtained a new constant velocity. Due to scale-equivalence, spacetime for an observer in the moving frame would then still seem identical to the Minkowskian spacetime. Furthermore, the scale adjustment implied by the DIST process locally restores the Minkowskian line-element; we may think of this process as locally expanding the scale for the moving particle. However, this scale adjustment will also have the effect of *reducing the relative scale* of the original reference frame. This situation is a simple consequence of spacetime scale adjustment; locally increasing the metrical scale will, at the same time, reduce the relative scale of spacetime elsewhere. The implication is that the scales of accelerating particles appear contracted and that all frames in relative motion appear to have smaller metrical scales. Conversely, as seen from the moving frame, the scale of the reference frame appears contracted.

Existence in five-dimensional hyperspace

If the scale dimension is included by extending GR to five dimensions, the SEC universe may be seen as traveling in this new 5D world "at the speed of light". In the same sense as traveling in the 4D spacetime at the speed of light causes time to stand still, traveling as fast as possible in this new 5D world causes the scale to stand still. In other words, relative to an observer in the SEC the scale is always the same across the cosmos. Furthermore, like the speed of light diminishes distances to zero in the 4D spacetime, distances in this is new 5D world decreases to zero and everything that exists is in direct communication via the scale. This is consistent with the idea that all existence is non-locally connected and that a common cosmological time exists. The mathematical underpinning for this view may be found in Appendix V.

Comments on the Nature of Photons

What constitutes a photon? Is it a wave or a particle? The nature of a photon has always been mysterious to us and still is. Let me give you my own guess.

It is possible that the photon actually has a rest mass, m_0, although extremely small. If this were the case, there would be a corresponding Compton frequency, which as usual is given by

$$f_0 = \frac{m_0 c^2}{h}$$

If the mass is extremely small, this frequency will be very low, so we might visualize a very slow oscillation of the metrical scale, which would mean that a photon at rest could be a rather large particle with very low mass that in practice would be undetectable.

However, if this particle moves very close to the speed of light, it will by the SEC model appear to be contracted not only in its direction of motion but also perpendicular to its motion. If this were true the wavelike nature of a photon would be caused by its de Broglie matter-wave, which would have a wave-length that depends on its velocity. The closer it is to the "speed of light", which could be a maximal possible velocity, the shorter is its wave-length and the higher its kinetic energy. This implies that almost all the energy of a photon would be kinetic and, thus, that the energy of electromagnetic radiation is due to the velocity of photon particles all moving very close to "the speed of light".

The speed of light, c, could then be the highest possible velocity in nature, and all photons would move at slightly lower speeds. However, since the rest mass is extremely small, all photons must move very, very close to the limiting velocity, c. Furthermore, if a photon is a particle it will by the SEC model not only be contracted in scale but also be additionally contracted in the direction of motion by the factor $1/\gamma$. It will appear as if the Compton oscillation lies in a plane perpendicular to the motion. This agrees with the fact that electromagnetic waves oscillate in a plane perpendicular to their motions.

Note that this interpretation becomes possible if the scale of a moving particle contracts via the inertial scale factor. However by SR, this interpretation is not possible because no contraction occurs in directions perpendicular to the motion. Therefore, the mysterious nature of a photon might perhaps be blamed on special relativity, which makes it impossible to reconcile its particle and wave aspects.

New Knowledge and Inertial Field Energy

How come most innovators are young, or if no longer young, new to the field of study? I think it the might be because their thoughts and ideas are free to roam unfettered by prior knowledge and opinions. Here is quote:

> A MAN CANNOT BEGIN TO LEARN THAT WHICH HE THINKS HE ALREADY KNOWS.
>
> —EPICTETUS

The more you know, the harder it becomes to accept that the knowledge you have acquired over many years by great effort actually might be wrong. This is particularly true if you depend on this knowledge in your profession and daily life. Knowledge is a double-edged sword; too much of might stifle innovation.

Often the most stunning advances come with an idea that at first may seem outlandishly crazy—an idea that may not be put forward by a learned individual because it could become an embarrassment. Most "new" ideas have already been considered and rejected in the past. However, times change, and what in the past seemed impossible might now have become possible. But, if you once have learned that something is impossible you will seldom reconsider it.

And we all "know" that perpetual motion is impossible, don't we? We know this based on classical physics, which in turn is based on the assumption that the pace of time has always been the same. But now with the SEC model, the scale of time gradually expands, which makes perpetual motion possible not only cosmologically but in principle even locally. If acceleration changes the scale of spacetime, as I have suggested, there is a corresponding additional previously unknown form of energy, Inertial Field Energy, contained in the dynamically changing scale of spacetime. It is given by the energy-momentum tensor of GR. This is discussed in my paper [Masreliez, 2009], which suggests that it might be possible to generate *negative field energy* by acceleration profiles satisfying $DIV(a) < 0$ where $DIV(\ )$ is the mathematical divergence operator. Hence it might be possible to reduce, or perhaps even eliminate, the inertial mass.

What previously was impossible may now become possible.

The Old and New Worldviews

Typically we visualize the world as a stage upon which all activity and all existence take place. We make our entry onto this stage when we are born, we live our lives, and we exit it when we die. We believe that people in the past inhabited the same old world as we do and that the difference between the past and the present merely is a distance in time. This view has not changed for millennia; people still think in terms of occupying the same old world as their predecessors did.

The new worldview is drastically different in that the world as it exists right now is not the same as the world of the past. Since the scale of space and time keeps changing, the present world differs from the world of the past.

The changing cosmological scale may be crucial to all existence. Everything is sustained by the dynamically changing scale; life would not exist without it. As humans, we are all "surfing on the wave of the expanding scale" that keeps everything moving toward the future— from electrons in atoms in our bodies, to life processes on Earth, to radiating Suns, to galactic upheavals and cosmological movements. We exist here and now, but we are not in the same world as just a moment ago. The past is gone forever, and we are in different world now with a different scale.

Truth, Innovation, and Status Quo

There is a catch-22 in the advancement of knowledge: on one hand, one must become familiar with what previously has been discovered, but on the other hand, one should also be skeptical of this previous knowledge. This seems like a contradiction, but it is imperative for the advancement of science and human knowledge in general that we absorb knowledge with a skeptical eye. Unfortunately, some of us confuse established scientific epistemology with what is true; they believe that what is documented and generally accepted is the actual and uncontestable truth.

This misunderstanding is not that unusual in the teaching profession where known epistemology often is taught as being the absolute truth. However, we must realize that nobody knows the "truth." In fact, we must realize that the truth will always be elusive, because it changes with the passage of time and our increasing level of awareness. Life would be much simpler if we always knew what is true and what is false. However, if we already knew this, no advancement would be possible, because new and unfamiliar ideas would then always be wrong merely because they did not comply with the known truth.

Looking back at the scientific development during the past century, we find that Einstein's relativity theories that initially were tentative now have become the established truth (although Einstein did not receive the Nobel Prize for them). Challenging this now firmly established epistemology may be seen as a personal attack on Einstein, who has assumed the status of guru. However, this attitude does not help scientific innovation. In fact, it obstructs innovation and hinders the development of science.

Perhaps the reader remembers that Einstein once was a rebel who said,

"If one studies too zealously, one easily loses his pants."

And that Richard Feynman said,

"Science alone of all the subjects contains within itself the lesson of the danger of belief in the infallibility of the greatest teachers in the preceding generation. Learn from science that you must doubt the experts. As a matter of fact, I can also define science another way: Science is the belief in the ignorance of experts."

It is well-known that blindly adhering to already known facts and procedures will hinder the implementation of new ideas. Unfortunately this means that really new and important ideas seldom originate in large research organizations.

The possibility that a dynamic spacetime scale may play a major role in all existence is something totally new, well beyond currently known epistemology. Consequently it has not yet been accepted; it has not yet become part of science.

Science and Passion

Anyone who thinks science is devoid of passion is wrong. The yearning for truth and beauty that lures many into a life of science is similar to the love of art, music, or poetry. The feeling is also similar to falling in love; sometimes it may become a passion that burns with a strong flame, difficult to explain to those who never have experienced it. The joy of discovery lifts the spirit into higher realms with a feeling of making contact with something higher and spiritual, something ethereal and beautiful. Maybe it is similar to religious revelation.

Although the pursuit of truth and beauty has been both my motivation and my reward, I would, of course, be happy if my efforts over all these years would turn out to have some lasting value and perhaps contribute to our understanding of the world. But, above all, I would like others to experience the delight and awe I have felt.

I am 73 years old now, and since it usually takes quite some time before a revolutionary new idea like the SEC model with its DIST process gains acceptance, recognition might not come in my lifetime. However, I want to tell you about the new world I have discovered. Like Marco Polo, I have visited a new land and want you to learn about it and feel what I feel.

The problem with something really new and different like this is that the scientific community is totally unprepared for it; *it is blindsided*. Even if the new theory makes both logical and scientific sense, people in science might reject it offhand if its ramifications and implications require major reassessment of scientific epistemology and philosophy. Many will not be inclined to deal with this can of worms, particularly if it threatens their personal legacies or perhaps even their livelihoods.

Furthermore, since the science community decides what is to be considered "science" and where the research money should go, work on the new heretic ideas presented in this book may not be funded for some time.

Leo Tolstoy says it better than I can:

I KNOW THAT MOST MEN, INCLUDING THOSE AT EASE WITH PROBLEMS OF THE GREATEST COMPLEXITY, CAN SELDOM ACCEPT EVEN THE SIMPLEST AND MOST OBVIOUS TRUTH IF IT BE SUCH AS WOULD OBLIGE THEM TO ADMIT THE FALSITY OF

CONCLUSIONSWHICHTHEYHAVEDELIGHTEDINEXPLAINING
TO COLLEAGUES, WHICH THEY HAVE PROUDLY TAUGHT
TO OTHERS, AND WHICH THEY HAVE WOVEN, THREAD BY
THREAD, INTO THE FABRIC OF THEIR LIVES.

—LEO TOLSTOY

When I was a young boy, my mother once told me the story about Heinrich Schliemann, who found the site of the ancient city of Troy. After a successful business career, he was able to pursue a dream he had had ever since he was 8 years old, when his father took him on his knee and told him the story about the Iliad and the city of Troy. At the age of 47, he was ready to start searching for the site of Troy, which he later found with the help of the British archeologist Frank Calvert. This story appealed a lot to me, and I dreamt of being able to do something like that myself when I grew up. Since I always have had a knack for physics, I was happy to be able to pursue the research presented in this book after selling our little family business in 1995, thus fulfilling a childhood dream.

This research has been rewarding for me with many exciting and enthralling discoveries. I now live in a different and better world than the one I once lived in before this research began. I am happy in my new world, because living in the old world was gloomy, which is well expressed by the Nobel laureate Steven Weinberg:

"The more the universe seems comprehensible, the more it also seems pointless."

The SEC theory offers a bright, new, and better world.

A Bull in a China Shop

In looking back on what I have been doing during the past 17 years I find that I unwittingly might have, like a bull in a china shop, shattered cherished beliefs by challenging Newton's first law of motion, invalidating Einstein's SR theory, and modifying his GR theory. And the planets are now falling toward the Sun!

Of, course I might be wrong, and many probably hope I am. However, should I turn out to be right, it would in no way lessen the standings of the great spirits of science of the past, who gave us the base for future exploration. Without GR, the importance of dynamic spacetime metrics would not have been found!

I did not plan these consequences; I merely followed where the scale-expansion idea led me. The discovery of the link between GR and QT came as a surprise to me as did the explanation to the inertial force. In fact, about five years ago (2005) I remember saying that I left an explanation to inertia to future generations. Had I been in academia, I would probably have given up on this pursuit a long time ago to avoid being ostracized.

It is possible that we will look back at this development with different eyes in the future. Perhaps Newton's first law and Einstein's two relativity theories will be seen as being different branches on the same old tree, a tree grown under the constraint of a fixed spacetime scale and a fixed pace of time. Quantum theory is an estranged branch that really never has belonged to traditional physics. Being a theory that unknowingly deals with a dynamic spacetime metrics, it seemed strange and unfamiliar when it was first was introduced in the early 1900s. With the SEC theory, this unnatural separation ends, the dynamic spacetime scale being the great unifier.

But Can All This Really Be Right?

The reader who has had the tenacity to stay with me up to this point might naturally wonder if these revolutionary new ideas and explanations really can be true. I believe that we shouldn't think of this new physics as being true or false, but should simply ask ourselves if it provides a better description for our world.

Undoubtedly the findings presented in this book will be revised in the future; this is always the case with new ideas. For example, I modeled the DIST process using GR simply because GR represents our current level of knowledge. If an alternative to GR had been available, I would probably have used this instead. In other words, the DIST loop of incremental scale adjustments is merely a way to describe the new scale-expansion process in the terms of known 4D physics. It is very likely that a new and better representation will become available in the future. In fact, the DIST process may also be modeled by introducing a fifth dimension in GR, as in the Kaluza-Klein theory.

In assessing the new developments of this book, it appears likely that dynamic spacetime metrics are of great importance regardless how they are modeled, because the changing scale makes use of the fundamental cosmological symmetry of scale-equivalence, which has not been properly recognized in the past. And, since it explains many seemingly diverse and previously unexplainable features of the universe, it simply cannot be without merit.

Obstruction of the Preconceived

I hope that the reader by now has gained some appreciation of the potential implications of a dynamic spacetime scale. It could resolve a whole array of problems plaguing contemporary physics, as primarily represented by Einstein's two relativity theories and quantum mechanics. Although both of his theories go as far as possible within the four dimensions of space and time, they do not suffice to describe the world.

In retrospect, and with 20/20 hindsight, it might seem strange that physics in the past overlooked the most fundamental cosmological symmetry of all: scale equivalence. But perhaps it is not so strange after all; often an implicit presumption that we all share about our existence obstructs development, because we do not see it. In our short life span, the very slow cosmological expansion is not directly noticeable. The proposition that the scale of everything, including ourselves, is steadily changing seems farfetched if not crazy.

Only by analyzing the consequences of cosmological scale-expansion and comparing its predictions with observations can we begin to realize that the scale-expansion idea might not be so crazy after all. But, unfortunately it is impossible to make this analysis by strictly adhering to known physics because GR cannot model a dynamic scale as it is perceived by the co-expanding observers we all are.

We are caught in a vicious catch-22: In order to become acceptable, a "scientific" cosmos model must use known physics. But, if the universe cannot be modeled by known physics, the cosmos must be modeled by unknown physics, which is unacceptable.

Therefore, what is acceptable is invalid, and what is valid is unacceptable.

I am attempting to break out of this vicious loop by proposing that known physics should be revised so that it can accommodate a dynamic spacetime scale.

We are facing a scientific revolution that must raze the old in order to make place for the new.

CHAPTER 11

The Missing Dimension

SOMETIMES IT'S NECESSARY TO GO A LONG DISTANCE OUT
OF THE WAY IN ORDER TO COME BACK A SHORT DISTANCE
CORRECTLY.
—EDWARD ALBEE

Physics as we currently know it may be inadequate. We know that something important surely must be missing, since we do not even know what is causing the progression of time, which is the most important physical aspect of our existence. With a few notable exceptions nobody has seriously pondered this puzzling question, and those who have thought about it have usually abandoned this line of inquiry after having found no explanation. Since in the past no answer ever has been found to the progression of time, some believe that an explanation is beyond human reach.

The same holds true with the question of the creation of the universe; I think that to most of us the creation poses an irresolvable problem even in the Big Bang scenario. Yet, based on our current understanding as given by general relativity, scientists have concluded that the universe was created about 14 billion years ago. It is surprising that this actually is the currently accepted belief. Given the impossibility of creating anything at all from nothingness, something surely must be missing in our current understanding.

With this book you have found that general relativity actually is incomplete since it cannot model a dynamic spacetime scale as it would be experienced by a co-moving observer. This could be the main reason why an explanation to the progression of time has been missing: it could be caused by the cosmological scale-expansion. And this would also resolve the creation puzzle because cosmological scale-expansion allows perpetual existence. To those who find eternal existence mindboggling, I think that the creation of everything from nothingness in an incredibly short time is even more mind-boggling! It cannot be explained by known physics, and it can never be confirmed. Therefore, in my opinion it is not science. Yet, this is just what the Big Bang model claims!

In retrospect we may find that ever since the dawn of modern physics with giants like Galileo Galilee and Isaac Newton, we may have overlooked a fundamental aspect of existence—the dynamic spacetime scale. We have always believed that we live in a world of time and space, and that all motion takes place as changing locations in time and space. But we now find that this might not be the whole story; there may be an even more important aspect of motion in the form of a dynamic scale of space and time that participates in all motion.

Of course, as soon as we realize that the scale of space-time might change with motion, it seems natural that this new degree of freedom ought to exist, because there does not seem to be any sufficient reason why the scale of material objects should be predetermined and always remain the same. What could possibly determine it? Perhaps we should not blame people in the past for overlooking this possibility because they were convinced that God had created the world and also determined the scale of things.

An unsubstantiated, firmly held preconception supported by religious dogma may have kept us from discovering the truth.

And now in our supposedly more enlightened times, we have simply gone along with this outdated thinking without even considering the possibility that the scale of spacetime may change.

Taking into account the missing "dimension" in the form a dynamic spacetime scale would not merely imply a minor adjustment to our current understanding, because this new previously neglected degree of freedom is arguably more important than both space and time! Whereas space and time may be seen as the "stage" upon which existence plays out, the dynamic scale is what keeps the cosmos going. It could be the perpetual energy source of the universe. Without the cosmological scale-expansion, time would not progress, there would be no energy; nothing would exist. It induces a cosmological energy that defines the spacetime geometry. The expanding scale keeps everything in motion; it could be the eternal life-giving "Sun" of the cosmos.

The cosmological scale-expansion may also implicitly determine the scale of things because it could induce oscillations in the spacetime scale that could be the source of both matter-energy and electromagnetic energy. And this oscillation could determine the pace of atomic time, which could be a temporal reference valid across the cosmos. The fact that this temporal reference exists is a fundamental property of nature that cannot be overemphasized or denied because the structure of atoms and matter is determined by this cosmological clock. It provides a cosmological reference of existence that together with a constant velocity reference, the speed of light, determines the structure of particles and the scale of all existence. This disagrees with current opinion, which does not recognize a cosmological temporal reference, but you have seen that this might be a misunderstanding caused by special relativity with its relativistic time. Again, although time is relative *as perceived* in moving frames, locally clocks may always run at the same pace.

We may find that the expanding spacetime scale is by far the most important physical aspect of our lives, yet science has in the past been oblivious of its existence! Realizing its significance, we may find that most of what we think we know about the world today is false. It will not suffice to merely amend our current epistemology by adding the new dynamic scale aspect. We have to start over from the very beginning and carefully reconstruct our worldview beginning with Newton's laws of motion, since his first law no longer holds in its original form. This could be a watershed in human development that will revolutionize our worldview.

In order to grasp the enormity of the situation, let's consider the development of the theory of motion that began with Newton's celebrated laws of motion. A huge amount of work over several hundred years by thousands of people was based on the assumption that Newton's laws are absolutely true and that all motion takes place in space and time. The dynamic scale was unknown. How much of this imposing edifice would remain intact after taking into account the dynamic scale? For example, thermodynamics is based on the assumption that the scale (pace) of time has always remained the same, which leads to the conclusion that perpetual motion is impossible. Yet the cosmos could be the prime example of perpetual motion, so the laws of thermodynamics fail at a most fundamental level. What else will fail? It is possible that most of our current knowledge will be proven incomplete or simply wrong.

This is the downside of the new ideas presented in this book. However, the upside is overwhelmingly positive: we may discover a new world with new possibilities that previously were deemed impossible. This new world will seem as magical to us as our present world would have seemed to people a few hundred years ago.

Bootstrapping existence

We have always wondered about the origin of the world; we have wondered how it began and how it will end. But, posing this question might be a mistake due to our short lifespan and limited insight, because it now appears likely that the world has no beginning or end. Existence may be perpetual.

Logically there can have been no beginning because if there were it would leave the unanswerable question: "What happened before the beginning?" Ancient civilizations have repeatedly faced the same conundrum often "resolved" by creation myths featuring otherworldly deities.

A well-known scientist (some say it was Bertrand Russell) once gave a public lecture on astronomy. At the end of the lecture, a little old lady at the back of the room got up and said: "What you have told us is rubbish. The world is really a flat plate supported on the back of a giant tortoise." The scientist gave a superior smile before replying, "What is the tortoise standing on?" "You're very clever, young man, very clever," said the old lady. "But its turtles all the way down!" http://en.wikipedia.org/wiki/Turtles_all_the_way_down

Of course, this is quite humorous until one realizes that any creation scenario no matter how fancy it might be will face the same "turtles all the way down" problem, including the Big Bang theory.

No, we have to face the fact that existence may be perpetual.

I have earlier in the text mentioned that when Einstein first saw Karl Schwarzschild's solution to his GR field equations, he found a potential problem; it implied that infinitely far away from matter the geometry of spacetime is flat and Minkowskian with disappearing spacetime energy density. He argued that if the right hand side of the GR equations that represent energy density disappears, the left hand side representing geometry should also disappear. In other words, without energy density the geometry of spacetime is undefined.

Therefore, there is something strange with Schwarzschild's solution. Like with SR, where the Twin Paradox was ignored, people have simply ignored this conceptual problem with GR. And, we have by this omission unknowingly wandered right into a dead end. Leading personalities in physics may not even be aware of the nature of the trap they now find themselves in.

However, the way out of this situation is quite simple as soon as one understands the nature of the problem and realizes that *Minkowskian spacetime cannot exist*. The reason is that without energy density there can be no spacetime geometry and no existence!

In the SEC model the scale-expansion paces out the progression of time and induces omnipresent cosmological energy density on the right hand side of the GR equations. The corresponding spacetime geometry appears on the left and side in the form of the SEC line-element.

We may say that space and time exists because their scale expands, and that this scale-expansion is made possible by the existence of spacetime. This is of course a "chicken-and-egg" situation whereby the cosmos pulls itself up by its bootstraps; existence is "bootstrapped". Some may object to this while yearning for some kind of "creation" of the world. However, whereas creation is physically and logically impossible the mutual dependence between spacetime geometry and its dynamic scale implied by the SEC model does not defy logic or physics, at least as we currently know them.

A final salvo

We exist because time progresses. And, time progresses because the cosmos expands. And, because the cosmos expands it induces spacetime energy that makes all existence possible. This understanding is fundamental to all knowledge, yet science has in the past been oblivious of it.

Famous leaders of science from Galileo to Einstein have in the past developed their theories while not knowing why or how time progresses, and today professors educating our young do not even mention the progression of time; it is not on the curriculum because nobody knows what is causing time to progress.

Yet, knowing the physical process the makes time progress will elevate humanity to a higher level of understanding. We will realize that there is more to the world than just the four spacetime dimensions, and learn how to tap into the ultimate energy source that powers the cosmos.

It's Up to You!

I am addressing this book to you who are seriously curious about the world we inhabit. You should ask yourself whether the SEC theory perhaps may be closer to the truth than the Big Bang creation myth. In doing so, you should realize that currently accepted physics simply cannot be the last word, because it cannot explain the progression of time. Therefore, it is not surprising that general relativity cannot model the universe, and as a consequence that the Big Bang theory simply is wrong.

Remember what Carl Sagan said:

"There are no sacred truths; all assumptions must be critically examined; arguments from authority are worthless."

Here is a quote from Buddha (Hindu Prince Gautama Siddharta, 563-483 B.C.):

"Do not believe in anything simply because you have heard it. Do not believe in anything simply because it is spoken and rumored by many. Do not believe in anything simply

because it is found written in your religious books. Do not believe in anything merely on the authority of your teachers and elders. Do not believe in traditions because they have been handed down for many generations. But after observation and analysis, when you find that anything agrees with reason and is conducive to the good and benefit of one and all, then accept it and live up to it."

There may be another and better explanation to our world than the currently accepted Big Bang model; a new cosmos model that better agrees with how we experience it.

So, when deciding what to believe, whether to remain trapped in the old world with an unexplainable beginning and a dismally gloomy end, or to embrace a new world of eternal existence with exciting new possibilities, I ask you to please independently make up your own mind.

WE ARE THE MUSIC-MAKERS
AND WE ARE THE DREAMERS OF DREAMS
WANDERING BY LONE SEA-BREAKERS
AND SITTING BY DESOLATE STREAMS;
WORLD LOSERS AND WORLD-FORSAKERS,
ON WHOM THE PALE MOON GLEAMS;
YET WE ARE THE MOVERS AND SHAKERS
OF THE WORLD FOREVER, IT SEEMS.
—ARTHUR O'SHAUGHNESSY

If you would like to stay in contact with the development of the SEC model and related subjects I invite you to join me at the SEC group www.estfound.org/progressionoftime

APPENDIX I

Deriving Cosmic Drag and Redshift

The geodesic relation for the SEC line-element yields [Masreliez, 1999]:

$$\beta = \frac{v}{c} = \frac{\beta_0 e^{-t/T}}{\sqrt{1 - \beta_0^2 + \beta_0^2 e^{-2t/T}}} \quad . \quad (AI.1)$$

Note that if the velocity of a particle initially equals the speed of light, it will always remain the speed of light:

$$\beta_0 = 1 \rightarrow \beta = 1 \text{ for all } t. \qquad (AI.2)$$

However, for small velocities this becomes:

$$\beta \ll 1 \rightarrow \beta \approx \beta_0 e^{-t/T} \rightarrow v \approx v_0 \cdot e^{-t/T} \qquad (A1.3)$$

This is the cosmic drag relation.

For velocities close to the speed of light,

$$\beta \approx 1 \rightarrow 1 - \beta^2 = \frac{1 - \beta_0^2}{1 - \beta_0^2 + \beta_0^2 e^{-2t/T}} \approx \left(1 - \beta_0^2\right) e^{2t/T} \quad \text{(AI.4)}$$

For these velocities close to the speed of light, the relativistic energy is then given by

$$E = \frac{E_{00}}{\sqrt{1 - \beta^2}} \approx \frac{E_{00}}{\sqrt{1 - \beta_0^2}} e^{-t/T} = E_0 \cdot e^{-t/T} \quad \text{(AI.5)}$$

If a photon is treated as a particle moving very close to the speed of light E_{00} might be interpreted as its "rest mass energy" and the velocity v_0 could be the local velocity of a photon of a certain spectral line.

A photon's energy diminishes with time and distance:

$$E = E_0 \cdot e^{-t \cdot T} = h \cdot f_0 \cdot e^{-t T}$$
$$f = f_0 \cdot e^{-t T} \quad \text{(A1.6)}$$

This is the redshift relation.

This treats the photon as a particle that moves very close to the speed of light with energy in proportion to the relativistic factor .

Theodor Kaluza showed that Maxwell's equations may be derived from a five-dimensional version of GR if the four off-diagonal metrical components of the fifth dimension correspond to the electromagnetic vector potential. However, there also is a scalar potential of unknown origin, which in the context of the SEC could model be the oscillating spacetime scale. This would be consistent with Oskar Klein's interpretation of the fifth dimension as being "curled up" and modeling quantum properties. This also suggests that the electromagnetic field might be a particular mode of metrical oscillation. Furthermore, as we saw, it proposes an ontological interpretation for a photon as being a packet of energy sustained by metrical spacetime oscillation. Almost all its energy could be relativistic, and the origin of its wave nature could be the de Broglie matter-wave resulting from its motion (see "The Nature of a Photon" in chapter 10).

APPENDIX II

Ephemeris Time and Universal Time

Ephemeris Time (ET) is based on the motion of the Earth around the Sun while Universal Time (UT) is based on the rotation of the Earth. UT is essentially the same as solar time. ET drifts positive at an accelerating rate and will in one century advance by about 30 seconds relative to UT. This difference between ET and UT is usually explained as being caused by a slowing rotation of the Earth caused by tidal braking action due to gravitational influences from the Moon and the Sun.

However, with the SEC theory there could be another explanation since this theory predicts that the angular motion of the Earth around the Sun accelerates in proportion to $\exp(t/T)$ with t=atomic time and that the rotation of the Earth slows down in proportion to $\exp(-t/T)$. The difference between ET and UT could therefore be interpreted as being partly caused by cosmic drag.

The Sun's acceleration due to cosmic drag causes ephemeris time to accelerate relative to atomic time. For small time intervals $t << T$ we have:

$$(ET - AT)_{sun} = T_{eph} - t \approx t + \frac{\Delta t^2}{2T} - t = \frac{\Delta t^2}{2T} \qquad \text{(AII.1)}$$

The spin-down of the Earth also contributes by:

$$(UT - AT)_{Earth} = -\frac{\Delta t^2}{2T} \qquad \text{(AII.2)}$$

Together this gives:

$$(ET - UT) = \frac{\Delta t^2}{T} \qquad \text{(AII.3)}$$

With $T = 14$ billion years this difference becomes 21 seconds in a century and with $T = 10$ billion years 30 seconds.

In addition the Earth's rotation might slow down due to tidal friction, which could account for the remaining difference between this estimate and the actually observed 30 seconds/cy.

This suggests that the difference between ET and UT mainly could be due to cosmic drag rather than tidal friction.

APPENDIX III

Explaining Spiral Galaxies, the Moon's Recession, and the Pioneer Anomaly

We saw in chapter 4 that spiral galaxies pose a major problem for the SCM. Their spiral arms and their flat rotation curves simply cannot be explained by known physics.

This appendix shows how both their spiral shape and flat rotation curves follow from the new physics suggested by the SEC model. It dives a bit deeper into the mathematics since it is impossible to explain what's going on in words. We will make use of GR in a derivation that demonstrates how the motion of a celestial body around a central gravitating mass might be a shallow inward spiral with uniform tangential velocity that does not depend on the radial distance to the center.

We start with the SEC line-element of GR, which in spherical coordinates with c=1 is [Masreliez, 1999]:

$$ds^2 = e^{2t/T}\left(dt^2 - dr^2 - r^2\left(d\theta^2 + \sin(\theta)^2 d\varphi^2\right)\right) \quad \text{(AIII.1)}$$

As before T is the Hubble time. This line-element explains how the Hubble time enters as a fundamental constant in the SEC model. Note that since $c=1$, formally T denotes both time and distance. Consider the coordinate transformation:

$$t' = T \cdot \cosh(\frac{r}{T}) \cdot e^{t/T} - T \quad \text{(AIII.2)}$$

$$r' = T \cdot \sinh(\frac{r}{T}) \cdot e^{t/T} \quad \text{(AIII.3)}$$

With this transformation the SEC line-element becomes:

$$ds^2 = dt'^2 - dr'^2 - r^2 \cdot e^{2t/T}\left(d\theta^2 + \sin(\theta)^2 d\varphi^2\right) \quad \text{(AIII.4)}$$

By GR's covariance this line-element is physically equivalent to (AIII.1). Here r and t in the last term may be found from the transformation relation (AIII.3).

We have from (AIII.3) using Taylor expansion:

$$r' \approx r\left(1 + \frac{1}{6}\left(\frac{r}{T}\right)^2 + \textbf{higher order terms}\right)e^{t/T} \quad \text{(AIII.5)}$$

The second term in the bracket is quite small even for something as large as a galaxy. At a radial distance of 100,000 light years we find with T=14 billion light years that the last term is smaller than 10^{-10} and may therefore safely be ignored. The corresponding transformation relation becomes:

$$r' = T \cdot \sinh(r/T) \cdot e^{t^T} \approx r \cdot e^{t^T} \qquad \text{(AIII.5)}$$

Using this in (AIII.4):

$$ds^2 \approx dt'^2 - dr'^2 - r'^2 \cdot (d\theta^2 + \sin(\theta)^2 \cdot d\varphi^2) \qquad \text{(AIII.6)}$$

This is the Minkowskian line-element for flat spacetime! Therefore, with these transformed coordinates the cosmological geometry is very close to Minkowskian and therefore all the classical laws of physics apply.

Similarly we get:

$$t' = T \cdot \cosh(r/T) \cdot e^{t^T} - T \approx T \cdot \left(e^{t^T} - 1\right) \approx t + \frac{t^2}{2T} \qquad \text{(AIII.7)}$$

Since classical physical relationships hold true with the transformed coordinates the conservation of angular momentum becomes for circular motion:

$$r' \cdot v' = r'^2 \cdot \omega' = \text{constant} = C \qquad \text{(AIII.8)}$$

However by the SEC theory there is loss of angular momentum:

$$r \cdot v = r^2 \cdot \omega = C \cdot e^{-t/T} \qquad \text{(AIII.9)}$$

The angular velocity is the time-derivative of the angular position:

$$\omega' = \frac{d\varphi}{dt'} = \frac{d\varphi}{dt} \frac{dt}{dt'} = \omega \cdot \frac{dt}{dt'} \qquad \text{(AIII.10)}$$

Relation (AIII.7) above yields:

$$dt' \approx dt \cdot e^{t/T}$$

$$\frac{dt}{dt'} \approx e^{-t/T} \qquad \text{(AIII.11)}$$

We find that:

$$\omega' = \omega \cdot \frac{dt}{dt'} \approx \omega \cdot e^{-t/T} \qquad \text{(AIII.12)}$$

Since ω' is constant:

$$\omega = \text{constant} \cdot e^{+t/T} \qquad \text{(AIII.13)}$$

In the SEC the angular velocity increases with time. Combining this with (AIII.9):

$$r = \text{constant} \cdot e^{-t/T} \qquad \text{(AIII.14)}$$

As modeled with the transformed coordinates, circular motion becomes inwardly spiraling motion with the SEC coordinates. However, the tangential velocity is still constant since:

$$v = \omega \cdot r = \text{constant}$$

Thus, according to the SEC model a celestial body orbiting around a central mass will slowly spiral inward with increasing angular velocity, but with constant tangential velocity regardless of its radial distance. This explains both the spiral galaxy arms and their flat rotation curves, providing strong support for the SEC model.

A more detailed analysis shows that the radial acceleration satisfies:

$$\frac{d^2 r}{dt^2} = -\frac{GM}{r^2} \cdot e^{-t/T} + r\omega^2 - \frac{1}{T}\frac{dr}{dt} \qquad \text{(AIII.15)}$$

The last term is cosmic drag.

The mysterious origin of the Moon was mentioned in chapter 5. It is estimated that the Moon recedes by 3.8 cm/year. If this is true the Moon must have been in contact with the Earth about 1.5 billion years ago, which we know is not the case. However, this estimate uses celestial mechanics based on Newton's laws of motion and gravitation and is therefore based on the transformed coordinates in relations AIII.2 and AIII.3. The relationship between the radial distance to the Moon expressed by these coordinates and the radial distance of the SEC model is given by AIII.5. Differentiating this relation with respect to time yields:

$$v' = \left(v + \frac{r}{T} \right) e^{t\,T} \qquad\qquad \text{(AIII.16)}$$

This relation implies that the rate of recession might be overestimated by r/T, which for $T = 10$ billion years is 3.8 cm/year (!) and for $T = 14$ billion year 2.7 cm/year. This shows that the estimate 3.8 cm/year might be wrong and that the Moon's recession rate could be much smaller or perhaps non-existent.

The SEC theory might resolve the enigmatic origin if the Moon. It might have been created at the same time as the Earth.

Chapter 5 also discusses the Pioneer anomaly, which is a difference in the space probe's velocity estimated by two different methods:

1. Direct measurement of the phase shift of a signal received from the space probe.

2. Measurements of the distance by ranging combined with trajectory (ephemeris) modeling.

By the first method the phase shift is directly measured using atomic time, which I will assume is the same as the SEC time t in the relations above.

The second method based on ephemeris modeling uses coordinates that are found by fitting the planetary ephemerides with assumption that Newton's laws apply. Therefore this method uses the transformed, primed, coordinates.

The measured phase shift is a combination of Doppler shift due to the outward motion of Pioneer 10 and the SEC model's redshift, which is present even in the solar system. This frequency redshift is given by:

$$\frac{f - f_0}{f_0} = e^{-\Delta t \cdot T} - 1 \approx -\frac{\Delta t}{T} \tag{AIII.17}$$

Here f is the received signal frequency and f_0 the by the space probe transmitted frequency. The time Δt is the signal transmission time between the probe and the Earth.

The second method estimates the velocity based on ranging data and then estimates the Doppler shift from this velocity. However, since it uses the transformed time instead of the SEC model's time (atomic time) it underestimates the frequency per relation (AIII.12) resulting in a positive discrepancy between the measured and estimated frequency due to the different time base:

$$\textbf{Redshift error due to time base} = \frac{\Delta t}{T} \tag{AIII.18}$$

This error cancels the SEC model's cosmological redshift (AIII.17). This is consistent with the fact that there is no cosmological redshift when using the transformed coordinates.

However, according to (AIII.16) the ephemeris modeling program also over-estimates the outward velocity by $u = r/T$, which gives an additional Doppler shift error:

$$-\frac{u}{c} = -\frac{r}{cT} = -\frac{\Delta t}{T} \qquad \text{(AIII.19)}$$

The net result is a discrepancy in the estimated Doppler frequencies. If we believe that both estimates are correct there is an apparent acceleration that causes a Doppler shift discrepancy:

$$\frac{\Delta v}{c} = -\frac{\Delta t}{T}$$
$$\frac{\Delta v}{\Delta t} = -\frac{c}{T} \qquad \text{(AIII.20)}$$

It appears as if the outward motion is being slowed down by a mysterious inwardly pointing acceleration c/T. This agrees well with the actually observed Pioneer Anomaly.

Therefore, the SEC theory might also explain the Pioneer Anomaly.

The Pioneer Anomaly could be a direct verification of the SEC model based on measurements in the solar system. Although optical observations also have confirmed orbital secular drifts of the planets, which could be caused by the use of different spacetime coordinates; these discordances are currently treated as being "unexplainable". They persist in spite of improving the modeling by taking into account more than 300 asteroids. In practice this discordance is overcome by simply discarding older observations that no longer agree with the more recent ephemerides.

The fact that three seemingly unrelated discordances may be explained as being caused by the same systematic modeling error should provide strong support for the SEC model.

In the past we have lived in a world where the background geometry of spacetime was believed to be flat and Minkowskian, and we have based all celestial mechanics on this assumption. However, the cosmological background spacetime might actually be curved by the cosmological scale-expansion and the traditional celestial mechanics might not apply.

This would imply a huge paradigm shift.

APPENDIX IV

Deriving Quantum Mechanics from General Relativity

At first I did not intend to include this appendix in the book since it is rather technical. However, the connection between General Relativity and Quantum Mechanics (QM) has remained elusive for a long time and finding it could be of considerable general interest. In this appendix I will use QM instead of QT to denote the theory based on the Schrödinger equation (and Heisenberg's matrix approach).

The fact that the link between GR and QM has been missing ever since the beginning of quantum theory suggests that this connection perhaps cannot be found within the four-dimensional spacetime of GR. However, this is not true because the derivation below is made using GR. In other words, the derivation does not use knowledge beyond current physics except in one respect; *it assumes that the metrical scale of spacetime for a particle oscillates.*

Although an inhabitant in the SEC may not locally experience the scale-expansion on a macroscopic level, the DIST process suggests that the scale of spacetime may oscillate, and this possibility became the starting point for this investigation. As you will see it leads to a link between GR and QM.

It requires the following three assumptions:

A1: The 4D scale of a particle confined to a small volume oscillates at the Compton frequency.

A2. The geometry of a particle in motion may be modeled by the Minkowskian line-element modulated by an oscillating scale at the Compton frequency.

A3. The linear part of the Ricci scalar of GR for this oscillating line-element disappears.

I will show that these three assumptions allow QM to be derived from GR.

Consider the Minkowskian line-element with $c = 1$ modulated by a dynamic scale:

$$ds^2 = e^{2g(t,x,y,z)} \left(dt^2 - dx^2 - dy^2 - dz^2 \right) \qquad \text{(AIV.1)}$$

Let us assume that the function $g(\cdot)$ may be factored into a spatial and an oscillating temporal part:

$$ds^2 = e^{\text{Re}(2C \cdot h(x,y,z)) \cdot e^{-i\omega t}} \left(dt^2 - dx^2 - dy^2 - dz^2 \right) \text{(AIV.2)}$$

The use of a complex exponent is to be interpreted as the real part, for example $exp(-iw\,t)$ means $cos(w\,t)$ and $i \cdot exp(-iw\,t)$ means $sin(w\,t)$. The reason we can do this is that all relations derived in the following are linear so that their real and imaginary parts may be separated. The complex exponential also simplifies the derivation and leads to results familiar from QM. In the following the label Re( ) is omitted.

Since the following derivation formally uses standard differential methods in 4D spacetime I will use the Lorentz transformation instead of Voigt's transformation (combined with scale adjustments), because SR with its Lorentz transformation is the best we can do in four dimensions. Also, this will demonstrate that a link between GR and QM exists with currently known physics.

With these preliminaries let us first consider motion of a spatially confined volume modeled by the line-element (AIV.2). With constant velocity v in the x-direction the Lorentz transformation is with $c = 1$:

$$x = \gamma(x' - vt')$$
$$t = \gamma(t' - vx')$$
$$\gamma = \frac{1}{\sqrt{1 - v^2}}$$

(AIV.3)

The modulating part of the exponent in the metric then becomes:

$$2Ce^{-i\omega t} \rightarrow 2Ce^{i\gamma\omega(vx' - t')}$$

(AIV.4)

If the modulation is confined to a particle the spatial modulation in (AIV.4) is reminiscent of the quantum mechanical wave function of a moving particle with wave number:

$$k = \gamma\omega v$$

(AIV.5)

Thus, motion of a spatial volume with oscillating metrics has the effect of spatially modulating the phase of this oscillation.

Furthermore, every particle is associated with scale excitation at the *relativistic Compton frequency* given by:

$$m = \hbar\gamma\omega = \hbar\varpi$$
$$\varpi \triangleq \gamma\omega = 2\pi f \qquad \text{(AIV.6)}$$

Since $c = 1$ this relation is the familiar $E = mc^2 = hf$ where f is the relativistic Compton frequency.

The relationship between the momentum and the wave number is:

$$p = \hbar k = mv \qquad \text{(AIV.7)}$$

According to (AIV.4) motion would cause the Compton oscillation to become "phase modulated" and create a spatial wave, $exp(ikx)$.

Thus, if Compton oscillation in the scale of spacetime is associated with every particle this oscillation will during motion be accompanied by a metrical "matter-wave". This would provide support for de Broglie's two-wave idea. There is one "wave-particle" that could be the Compton oscillation associated with a particles motion in time, the other could be the deBroglie matter-wave associated with its motion in space. Both these quantum mechanical waves would then be modulations of the metrical scale of spacetime.

With this interpretation the quantum mechanical matter-wave is a relativistic phenomenon; it is a spatial wave in the metrical scale of spacetime induced by motion. Since the wave number of the matter-wave depends on the very high Compton frequency corresponding to a particle's matter energy, this small relativistic effect becomes significant even at relatively low velocities.

This interpretation would resolve the wave-particle duality since these two aspects are inseparable; the matter-wave is a direct consequence of the Compton oscillation and a particle's motion.

The previously mysterious fact that a particle behaves both as a wave and a particle, might find its natural explanation. Traditionally we think of a "particle" as something material an indivisible, but this might be wrong. Particles could be nothing but standing wave oscillations in the spacetime metrics that are sustained by the cosmological scale-expansion. Motion in time (and scale) might induce their Compton oscillation. This new understanding would also obsolete Bohr's Principle of Complementarity by explaining the dual wave-particle nature.

I mentioned in the text that Bohm and his followers have shown that a consistent quantum mechanical theory may be derived based on just three conditions:

C1. There exists a function, ψ (of unspecified ontology), which satisfies Schrödinger's wave equation.

C2. The motion of particles satisfies the relation:

$$p = \hbar \cdot Im \frac{\nabla \psi}{\psi} \tag{AIV.8}$$

C3. The motion is subjected to random disturbance.

A link between GR and QM will now be established by demonstrating that these three conditions may be derived from GR if the metrical scale oscillates. I will first show that the pilot function may be derived from the geodesic relation of GR.

Consider the scaled Minkowskian line element with a general dynamic scale function :

$$ds^2 = \phi^2(t, x, y, z) \cdot \left(dt^2 - dx^2 - dy^2 - dz^2\right) \quad \text{(AIV.9)}$$

The geodesic equation of GR is:

$$\frac{d^2x^\mu}{ds^2} + \Gamma^\mu_{\nu\lambda} \frac{dx^\nu}{ds} \frac{dx^\lambda}{ds} = 0 \quad \text{(AIV.10)}$$

For the x-coordinate this relation becomes with indices given by $x^0 = t$, $x^1 = x$, $x^2 = y$ and $x^3 = z$:

$$\frac{d^2x}{ds^2} = -\Gamma^1_{00}\left(\frac{dt}{ds}\right)^2 - \Gamma^1_{11}\left(\frac{dx}{ds}\right)^2 - \Gamma^1_{22}\left(\frac{dy}{ds}\right)^2 - $$

$$-\Gamma^1_{33}\left(\frac{dz}{ds}\right)^2 - 2\Gamma^1_{10}\left(\frac{dt}{ds}\right)\left(\frac{dx}{ds}\right) - 2\Gamma^1_{12}\left(\frac{dx}{ds}\right)\left(\frac{dy}{ds}\right) - 2\Gamma^1_{13}\left(\frac{dx}{ds}\right)\left(\frac{dz}{ds}\right) = $$

$$= -\Gamma^1_{00}\left(\frac{dt}{ds}\right)^2 + \Gamma^1_{11}\left(\frac{dx}{ds}\right)^2 - \Gamma^1_{22}\left(\frac{dy}{ds}\right)^2 - $$

$$-\Gamma^1_{33}\left(\frac{dz}{ds}\right)^2 - 2\left(\frac{dx}{ds}\right)\left[\Gamma^1_{10}\left(\frac{dt}{ds}\right) + \Gamma^1_{11}\left(\frac{dx}{ds}\right) + \Gamma^1_{12}\left(\frac{dy}{ds}\right) + \Gamma^1_{13}\left(\frac{dz}{ds}\right)\right] \quad \text{(AIV.11)}$$

We have:

$$\frac{d^2x}{ds^2} = \frac{d}{ds}\left(\frac{dx}{dt}\frac{dt}{ds}\right) = \left\lfloor \frac{d}{dt}\left(\frac{dx}{dt}\frac{dt}{ds}\right)\right\rfloor\left(\frac{dt}{ds}\right) = $$

$$\left(\frac{d^2x}{dt^2}\right)\left(\frac{dt}{ds}\right)^2 + \left(\frac{dx}{dt}\right)\left[\frac{d}{dt}\left(\frac{dt}{ds}\right)\right]\left(\frac{dt}{ds}\right) \quad \text{(AIV.12)}$$

From the line element (AI.1):

$$\frac{dt}{ds} = \frac{1}{\phi\sqrt{1-v^2}}; \text{ where } v = \sqrt{\dot{x}^2 + \dot{y}^2 + \dot{z}^2} \text{ and } \dot{x} = \frac{dx}{dt} \quad \text{(AIV.13)}$$

The bracket factor in the last term of (AI.4) therefore is:

$$\frac{d}{dt}\left(\frac{dt}{ds}\right) = -\frac{\dot{\phi}}{\phi^2\sqrt{1-v^2}} + \frac{v\cdot\dot{v}}{\phi(1-v^2)^{3/2}} = \left|-\frac{\dot{\phi}}{\phi} + \frac{v\cdot\dot{v}}{(1-v^2)}\right|\left(\frac{dt}{ds}\right)$$

$$\dot{\phi} = \frac{d\phi}{dt}$$

(AIV.14)

We may get rid of the dependence on s by dividing all terms in the geodesic by $(dt/ds)^2$. Rewriting the last term of (AIV12) using (AIV.14):

$$\dot{x}\left[\frac{d}{dt}\left(\frac{dt}{ds}\right)\right]\left(\frac{dt}{ds}\right) = \left[-\frac{\dot{x}\cdot\dot{\phi}}{\phi} + \frac{\dot{x}\cdot v\dot{v}}{(1-v^2)}\right]\left(\frac{dt}{ds}\right)^2 \quad \text{(AIV.15)}$$

The geodesic relation may now be written:

$$\left[\ddot{x} - \frac{\dot{x}\cdot\dot{\phi}}{\phi} + \frac{\dot{x}\cdot v\dot{v}}{(1-v^2)}\right]\left(\frac{dt}{ds}\right)^2 = \text{Right hand side of (AIV.11)} \quad \text{(AIV.16)}$$

The right hand side of (AIV.11) may also be written:

$$-\left[\begin{matrix}\Gamma^1_{00} - \Gamma^1_{11}\dot{x}^2 + \Gamma^1_{22}\dot{y}^2 + \Gamma^1_{33}\dot{z}^2 \\ +2\dot{x}\left\{\Gamma^1_{10} + \Gamma^1_{11}\dot{x} + \Gamma^1_{12}\dot{y} + \Gamma^1_{13}\dot{z}\right\}\end{matrix}\right]\left(\frac{dt}{ds}\right)^2$$

(AIV.17)

The Christoffel symbols are:

$$\Gamma^1_{00} = \Gamma^1_{11} = -\Gamma^1_{22} = -\Gamma^1_{33} = \frac{1}{\phi}\frac{\partial\phi}{\partial x}$$

$$\Gamma^1_{12} = \frac{1}{\phi}\frac{\partial\phi}{\partial y}; \qquad \Gamma^1_{13} = \frac{1}{\phi}\frac{\partial\phi}{\partial z}; \qquad \Gamma^1_{10} = \frac{1}{\phi}\frac{\partial\phi}{\partial t}$$

(AIV.18)

Substituting this into the bracket of (AIV.17):

$$-\frac{1}{\phi}\left[\begin{array}{l}\dfrac{\partial\phi}{\partial x}-\dfrac{\partial\phi}{\partial x}\left(\dot{x}^2+\dot{y}^2+\dot{z}^2\right)\\+2\dot{x}\left(\dfrac{\partial\phi}{\partial t}+\dfrac{\partial\phi}{\partial x}\dot{x}+\dfrac{\partial\phi}{\partial y}\dot{y}+\dfrac{\partial\phi}{\partial z}\dot{z}\right)\end{array}\right]= \tag{AIV.19}$$

$$-\frac{1}{\phi}\left[\frac{\partial\phi}{\partial x}\left(1-v^2\right)+2\dot{x}\dot{\phi}\right]$$

Together with (AIV.16) we get:

$$\ddot{x}+\frac{\dot{x}\cdot v\dot{v}}{\left(1-v^2\right)}=-\frac{1}{\phi}\left[\frac{\partial\phi}{\partial x}\left(1-v^2\right)+\dot{x}\dot{\phi}\right] \tag{AIV.20a}$$

Similarly for the other two components:

$$\ddot{y}+\frac{\dot{y}\cdot v\dot{v}}{\left(1-v^2\right)}=-\frac{1}{\phi}\left[\frac{\partial\phi}{\partial y}\left(1-v^2\right)+\dot{y}\dot{\phi}\right] \tag{AIV.20b}$$

$$\ddot{z}+\frac{\dot{z}\cdot v\dot{v}}{\left(1-v^2\right)}=-\frac{1}{\phi}\left[\frac{\partial\phi}{\partial z}\left(1-v^2\right)+\dot{z}\dot{\phi}\right] \tag{AIV.20c}$$

Combining these we get in vector notations:

$$\mathbf{v}=\left(\dot{x},\dot{y},\dot{z}\right)$$
$$\dot{\mathbf{v}}+\frac{\mathbf{v}\cdot v\dot{v}}{\left(1-v^2\right)}=-\frac{1}{\phi}\left[\nabla\phi\left(1-v^2\right)+\dot{\phi}\mathbf{v}\right] \tag{AIV.21a}$$

Reintroducing c:

$$\dot{\mathbf{v}}+\frac{\mathbf{v}\cdot v\dot{v}}{\left(c^2-v^2\right)}=-\frac{1}{\phi}\left[\nabla\phi\left(c^2-v^2\right)+\dot{\phi}\mathbf{v}\right] \tag{AIV.21b}$$

Now consider the scale function:

$$\phi = e^{C \cdot h \cdot \exp(-i \cdot \omega \cdot t)}$$

(AIV.22)

We get with c=1:

$$\dot{\mathbf{v}} + \frac{\mathbf{v} \cdot \dot{\nu}}{\left(1-v^2\right)} = -C \cdot h \cdot e^{-i\omega t}\left[\frac{\nabla h}{h}\left(1-v^2\right)+\left(\frac{\dot{h}}{h}-i\omega\right)\mathbf{v}\right]$$

(AIV.23)

The rapid modulation of the phase with changing velocity implied by the imaginary term within the bracket disappears if:

$$\omega\mathbf{v} = \mathrm{Im}\left[\frac{\nabla h}{h}\left(1-v^2\right)+\frac{\dot{h}}{h}\,\mathbf{v}\right]$$

(AIV.24)

$$m\mathbf{v} = \mathbf{p} = \hbar \cdot \mathrm{Im}\left[\frac{\nabla h}{h}\left(1-v^2\right)+\frac{\dot{h}}{h}\mathbf{v}\right]$$

(AIV.25a)

This is the relativistic version of the deBroglie-Bohm pilot wave function.

The last term in the bracket is very small if v<<c and we then get the usual deBroglie-Bohm pilot wave function:

$$m\mathbf{v} = \mathbf{p} \approx \hbar \cdot \mathrm{Im}\left[\frac{\nabla h}{h}\right]$$

(AIV.25b)

We also have:

$$\dot{\mathbf{v}} + \frac{\mathbf{v} \cdot \dot{\nu}}{\left(1-v^2\right)} = -C \cdot h \cdot e^{-i\omega t} \cdot \mathrm{Re}\left[\frac{\nabla h}{h}\left(1-v^2\right)+\frac{\dot{h}}{h} \cdot \mathbf{v}\right]$$

(AIV.26)

According to the last relation there is cyclic acceleration excitation.

Example: We saw that for motion in the x-direction we have:

$h = e^{i\varpi xv}$;

$$p = \hbar \cdot \operatorname{Im}\left[\frac{\nabla h}{h}\left(1-v^2\right)+\frac{\dot{h}}{h}v\right] = \hbar \cdot \left[\varpi v\left(1-v^2\right)+\varpi \frac{dx}{dt}\cdot v^2\right] = \hbar\varpi v = mv$$

(AIV.27)

If the complex function h(x,y,z), which modulates the Compton oscillation, is proportional to the quantum mechanical wave function ψ, relation (AIV.25) is the de Broglie/Bohm momentum relation, i.e. the "pilot function"{Bohm, 1952}. Therefore the pilot function may be derived from GR's geodesic relation.

The two relations (AIV.25) and (AIV.26) speak volumes about the ontological nature of QM. If the metrical scale of a particle oscillates it will be subjected to cyclic disturbance that depends on the spatial scale function h that modulates the oscillation. And, the particle tends to move in the direction of increasing values for h. Also, this motion will disappear when the gradient of h disappears. The particle converges toward peaks of the wave function.

This provides an ontological explanation to the deBroglie-Bohm's guiding function; it may be derived directly from the geodesic equation of GR if the spacetime of a particle oscillates at the Compton frequency. Thus, the previously mysterious "pilot function" finds its physical explanation if a particle always is accompanied (and sustained) by oscillation of the spacetime metrical scale at the Compton frequency.

This fulfills conditions C2 and C3.

Next I will show that the Schrödinger equation also may be derived from GR.

According to assumption A4 the Ricci scalar for the line-element should disappear. This assumption is reasonable since it is satisfied if the energy-momentum tensor for vacuum disappears. (I will in this derivation ignore the small contribution from cosmological expansion.) A necessary (but not sufficient) condition for the Ricci scalar to disappear is a wave equation for the function g of (AIV.1) [Masreliez, 2005a]:

$$\Delta\big(g(t,x,y,z)\big) - \frac{\partial^2}{\partial t^2}\big(g(t,x,y,z)\big) = 0 \qquad \text{(AIV.28)}$$

Here is the Laplace operator. Consider the g-function:

$$g = C \cdot h(x,y,z) \cdot \exp[(-i(\omega + E/\hbar)t + i\omega\left(\oint (1+V/m) \cdot \boldsymbol{ds} \cdot \boldsymbol{n}\right)] \qquad \text{(AIV.29)}$$

I will as before assume that the temporal oscillation is at the Compton frequency, and that this oscillation is confined to a small spatial volume.

The corresponding line-element is:

$$ds^2 = \exp\left[2 \cdot C \cdot h(x,y,z) \cdot e^{-i(\omega+E)t/\hbar} \cdot e^{i\omega\oint(1+V)\cdot \boldsymbol{ds}\cdot\boldsymbol{n}}\right]\left(dt^2 - dx^2 - dy^2 - dz^2\right) \qquad \text{(AIV.30)}$$

In the line integral $\boldsymbol{ds}$ is a path increment vector and $\boldsymbol{n}$ a unity velocity vector corresponding to motion at the speed of light. This form may seem contrived but I will show that it leads to the Schrödinger equation. Let's analyze it.

The energy E is assumed to be constant and may be seen as giving a frequency adjustment to the Compton oscillation, while the potential function $V(x,y,z)$ adjusts the phase of the deBroglie matter-wave. We will assume both these influences are much smaller than the mass energy:

$$E \ll \hbar \varpi = m$$
$$V \ll m \tag{AIV.31}$$

The line integral corresponds to the deBroglie matter-wave; $\mathbf{ds \cdot n}$ corresponds to the product $\mathbf{x \cdot v}$ in (AIV.4). The line element (AIV.30) therefore models motion that is influenced by a variations of the Compton frequency given by E and phase modulations of the deBroglie mater-wave given by V.

Replacing the line integral with a sum of segments assuming that the vectors $\mathbf{n}_i$ on these segments are constant:

$$I(x,y,z) = \oint (1 + V/m) \cdot \mathbf{ds} \cdot \mathbf{n} =$$
$$= \sum_{x_0, y_0, z_0}^{x,y,z} (1 + V/m)(\Delta x_i \cdot n_{xi} + \Delta y_i \cdot n_{yi} + \Delta z_i \cdot n_{zi}) \tag{AIV.32}$$

If the intervals are small, differentiation of this sum with respect to x may be approximated by the contribution from the last term in the sum:

$$\frac{\partial I}{\partial x} \approx \frac{\Delta I(x,y,z)}{\Delta x} \approx (1 + V/m) \, n_x \tag{AIV.33}$$

After a second differentiation of the exponent in the integral and adding the contributions from y and z we get the term:

$$(1 + V/m)^2 \left(n_x^2 + n_y^2 + n_z^2 \right) = (1 + V/m)^2 \tag{AIV.34}$$

We therefore find, after carrying out the differentiations in (AIV.28) that the Ricci scalar disappears if the following two relations hold:

Terms not containing $\mathbf{n}_i$:

$$\nabla^2 h - \varpi^2 \left[\left(1 + \frac{V}{m} \right)^2 - \left(1 + \frac{E}{\hbar \varpi} \right)^2 \right] h = 0 \qquad \text{(AIV.35)}$$

Terms containing $\mathbf{n}_i$:

$$\varpi \left[2 \left(1 + \frac{V}{m} \right) \cdot \frac{\nabla h}{h} + \frac{\nabla V}{m} \right] \cdot \mathbf{n}_i = 0 \qquad \text{(AIV.36)}$$

Using (AIV.31) we have:

$$\left[1 + \frac{V}{m} \right]^2 = 1 + \frac{2V}{m} + \left(\frac{V}{m} \right)^2 \approx 1 + 2\frac{V}{m} \qquad \text{(AIV.37)}$$

$$\left[1 + \frac{E}{\varpi \hbar} \right]^2 = 1 + 2\frac{E}{m} + \left(\frac{E}{m} \right)^2 \approx 1 + 2\frac{E}{m} \qquad \text{(AIV.38)}$$

Substituting these in (AIV.35) we get the Schrödinger equation:

$$-\frac{\hbar^2}{2m} \nabla^2 h + (V - E) \cdot h = 0 \qquad \text{(AIV.39)}$$

This derivation may easily be generalized to the situation where the wave function h also depends on time. We then get the additional terms:

$$2i \left(\varpi + \frac{E}{\hbar} \right) \frac{\partial h}{\partial t} - \frac{\partial^2 h}{\partial t^2} \approx 2i\varpi \frac{\partial h}{\partial t} \qquad \text{(AIV.40)}$$

$$-\frac{\hbar^2}{2m} \left(2i\varpi \frac{\partial h}{\partial t} \right) = -i\hbar \frac{\partial h}{\partial t}$$

Moving this term to the right hand side of (AIV.36)

$$-\frac{\hbar^2}{2m} \nabla^2 h + (V - E) \cdot h = i\hbar \frac{\partial h}{\partial t} \qquad \text{(AIV.41)}$$

This is the time dependent Schrödinger equation.

A similar derivation of the Schrödinger equation for the electromagnetic field may be found in [Masreliez, 2005].

Relation (AIV.32) does not depend on the velocity vectors n_i. If we associate these vectors with a trajectory it means that the Schrödinger equation applies regardless of a particle's motion. Furthermore, if relation (AIV.36) is satisfied for n_i it is also satisfied for $-n_i$. Therefore, it implies that the Schrödinger equation applies even for a particle "at rest" subjected to back and forth motion. In other words, the mere presence of an oscillating volume (particle) at some location creates a response from its environment given by the wave functions of QM, which could be modulations of the metrics of spacetime. This may influence the subsequent motion of the particle, and since this influence takes place via the metrics it could be non-local, allowing instantaneous influences independent of separation distance.

In other words, Schrödinger equation does not model motion but models resonance conditions in the metrics of spacetime that depend on geometry, energies, and fields.

This development shows that if the scale of spacetime oscillates at the Compton frequency the Schrödinger equation is a necessary condition for the disappearance of the Ricci scalar of GR. The finding that the deBroglie-Bohm pilot function and the Schrödinger equation both may be derived from GR and that there also is random influence implies that Bohm's conditions C1, C2 and C3 are all satisfied.

In other words, QM may be derived from GR.

Furthermore, it suggests that the quantum mechanical wave functions may have physical meaning; they could correspond to modulation of the Compton oscillation of the scale of spacetime.

This would allow us to merge GR and QM into a single more complete theory, ending their century-long estrangement. The probabilistic interpretation of QM would be abandoned in favor of new physics based on dynamic spacetime metrics. The behavior of the quantum world would no longer be something mysterious and probabilistic but would be a consequence of influences via the dynamic scale of spacetime.

Although you may appreciate this ontological explanation to quantum mechanics it must be admitted that it currently does not address several aspects of the quantum world, for example "spin", and it may therefore be ignored by mainstream experts who in the spirit of the Copenhagen school consider an ontological explanation unnecessary or even undesirable. But, like the epicycles of the past described the motions of the planets without giving any answer to the question "why", the currently popular purely mathematical and probabilistic approach to quantum theory (including string theory) may not contribute as much to our understanding of the world as even a simplistic ontological explanation will. When Copernicus presented his moving Earth model it did not model the motions of the planets with the same accuracy as the epicycles. However, it still became the preferred explanation because of its simplicity. This book may be the beginning to deeper understanding of the quantum world to be developed in the future.

I think Einstein was right; God does not play dice.

APPENDIX V

The Cosmological Expansion as Motion in Five-dimensional Hyperspace

An interesting special case of the canonical line-element is with $c=1$:

$$ds^2 = u^2(dt^2 - dx^2 - dy^2 - dz^2) - (L \cdot du)^2 \qquad \text{(AV.1)}$$

The first term corresponds to 4D spacetime. Like in the 4D spacetime of GR we may in this five-dimensional hyperspace let a "light-ray" geodesic (null geodesic) be given by $ds=0$. Since each observer is fixed in her local 4D frame we have $dx=dy=dz=0$ and therefore on this null geodesic:

$$u^2 dt^2 = L^2 du^2$$

$$u = e^{t/L} = e^{t/T} \qquad (AV.2)$$

Let L be the Hubble distance and $T = L/c$ be the Hubble time. The corresponding 4D line-element is:

$$ds^2 = e^{2t/T}(dt^2 - dx^2 - dy^2 - dz^2) \qquad (AV.3)$$

This is the line-element of the Scale Expanding Cosmos (SEC) model (Masreliez, 1999).

We find that the cosmological expansion may be seen as motion of dynamically scaled 4D Minkowskian space-times (associated with different observers) in 5D space with the spacetime scale as the fifth dimension.

We also find that:

$$dt = T \frac{du}{u} \qquad (AV.4)$$

In other words, cosmological scale-expansion could provide an ontological explanation to the progression of time. The Hubble time becomes a cosmological constant that relates the progression of time to a changing cosmological scale. As we saw, this fifth dimension also allows Inertia to be explained as resulting from a dynamic scale of 4D spacetimes embedded in 5D space.

It is interesting to investigate geodesic motion of the temporal coordinate in the hyperspace. The geodesic equation is:

$$\frac{d^2t}{ds^2} = -\Gamma^0_{04}\frac{dt}{ds}\frac{du}{ds} - \Gamma^0_{40}\frac{dt}{ds}\frac{du}{ds} = -\frac{2}{u}\frac{dt}{ds}\frac{du}{ds} \qquad (AV.5)$$

Here the index 4 corresponds to the scale u.

This may be integrated:

$$\ln(dt \,/\, ds) = -\ln(u^2) - \ln(C) = -\ln(Cu^2)$$

$$\frac{ds}{dt} = Cu^2 \tag{AV.6}$$

C is a constant of integration. From the canonic line element we get:

$$\left(\frac{ds}{dt}\right)^2 = u^2(1-v^2) - L^2\left(\frac{du}{dt}\right)^2 \tag{AV.7}$$

Consider the stationary case v=0. Using (AV.6) we get:

$$C^2 u^4 = u^2 - L^2\left(\frac{du}{dt}\right)^2$$

$$\frac{du}{u\sqrt{1-C^2u^2}} = \pm\frac{dt}{L} \tag{AV.8}$$

Note that replacing u by Cu in the line-element has no physical significance because of scale equivalence. We may choose C=1.

The inertial scale factor suggests a change in integration variable:

$$u = \sqrt{1-w^2} \tag{AV.9}$$

From which:

$$\frac{dw}{1-w^2} = \pm\frac{dt}{L} \tag{AV.10}$$

This may be integrated;

$$\sqrt{\frac{1+w}{1-w}} = e^{\pm\frac{t}{T}} \tag{AV.11}$$

In the SEC model the cosmological redshift is given by (Masreliez, 1999, 2004a). Since t<0 in the SEC model in the past:

$$z + 1 = e^{-\frac{t}{T}} \tag{AV.12}$$

With the corresponding cosmological distance modulus:

$$d = L \cdot \ln(z+1) \tag{AV.13}$$

Therefore:

$$\sqrt{\frac{1+w}{1-w}} = 1 + z \tag{AV.14}$$

This may be compared to the relativistic Doppler redshift, z, which with c=1 for a receding source is:

$$\sqrt{\frac{1+v}{1-v}} = 1 + z \tag{AV.15}$$

The cosmological redshift is related to the integration variable w as if it were caused by outward motion in space instead of being caused by "motion in scale" via the cosmological scale-expansion.
This derivation suggests that:

1. Although with cosmological scale-expansion there is no radial spatial motion, the observed redshift gives the impression that a radiating source is receding and that the redshift is due to Doppler shift.
2. The cosmological redshift may be due to an expanding spacetime scale and therefore be a purely geometrical effect induced without spatial motion.

3. The integration parameter w may be seen as corresponding to "inertial motion in scale". The "inertial scale factor" corresponding to this motion would then be given by relation (AV.9).

This demonstrates that geodesic motion of 4D spacetime in 5D hyperspace could explain cosmos as observed and experienced, and that there is symmetry between motion in scale and in space. Motion in general takes place in the metrical scale as well as in the four spacetime dimensions.

The scale of four-dimensional spacetime is an active cosmological degree of freedom that makes the world fundamentally five-dimensional. This suggests that the scale should be taken into account when modeling any kind of motion whether in space or time.

APPENDIX VI

Speculation on the Nature of Motion

In the beginning of this book I mentioned the mystery of motion and since the book's title is the progression of time, which is intimately connected with motion, it seems appropriate to return to this question here by the end of the book.

How does a particle move? Does it jump incrementally or does it change its dimensions and move like an inchworm? It turns out that appendices IV and V together with the explanation to the inertial force given in chapter 7 may offer a possible ontological explanation.

Consider again the 5D hyperspace line element:

$$ds^2 = u^2(dt^2 - dx^2 - dy^2 - dz^2) - (L \cdot du)^2 \qquad \text{(AVI.1)}$$

According to appendix V we may think of the 4D SEC cosmos as moving in this 5D hyperspace space "at the speed of light" with the fifth dimension, *u,* playing the role of "time". This line-element has two terms. The last term disappears if *u* is constant and the 5D line-element then collapses into the line-element for scale-equivalent 4D spacetime. On the other hand, motion in the 4D spacetime at the speed of light will cause the first term to disappear allowing motion purely in the fifth dimension, which may model scale transition. Spatial motion at the speed of light implies that time stands still in 4D spacetime, which makes the scale transition appear instantaneous.

This suggests that the incremental scale transition of the DIST process might be associated with motion in 4D spacetime at the speed of light.

Chapter 7 on inertia suggests that the spacetime scale for an accelerating particle contracts by the inertial factor *1-(v/c)²* and that relative to a co-accelerating observer this scale is incrementally "reset" via the DIST process to keep the line-element locally Minkowskian.

Consider the hyperspace line-element:

$$ds^2 = u^2 \cdot \left(1 - \left(\frac{v}{c}\right)^2\right)(dt^2 - dx^2 - dy^2 - dz^2) - (L \cdot du)^2 \quad \text{(AVI.2)}$$

We may think of acceleration as occurring in two steps:

- In the first step the scale contracts continuously via the inertial scale factor while the scale *u* remains constant *u= 1*. In this first step the world is 4D spacetime since the last term in (AVI.2) is zero. This step may be modeled by GR.

- In the second step there is spatial transition at the speed of light while the scale *u* adjusts *u= >(1-(v/c)²)⁻¹* thus "resetting" the 4D scale to one. In this second step, which corresponds to the discrete scale transition in the DIST process, the first term equals zero and the motion is solely in scale.

We find that motion might take place at the speed of light in spacetime together with transition in the fifth dimension.

This somewhat speculative ontological explanation would imply that motion takes place via transitions both in 4D spacetime and in five-dimensional hyperspace. The reader familiar with Richard Feynman's checkerboard approach to quantum theory may sense a connection here since he showed that random walks in space and time at the speed of light leads to Dirac's famous equation for the electron.

http://en.wikipedia.org/wiki/Feynman_checkerboard

However, I have not investigated this possible connection further.

When deriving the Schrödinger equation in appendix IV I expressed the increment in a line-integral as the scalar product of a displacement vector, **ds**, and a unit velocity vector, **n**, which corresponds to motion at the speed of light. I found that this allowed the Schrödinger equation to be derived from GR. This line-integral may be divided into number of short segments each modeling motion at the speed of light.

However, with 5D the line-element (AVI.2) we now find that incremental motion at the speed of light may be associated with changes in the fifth dimension that resets the dynamic scale and models the discrete scale adjustment of the DIST process. As we saw this scale adjustment appears to be instantaneous to an observer. It is therefore possible that particles always move in tiny steps at the speed of light combined with simultaneous scale adjustments, and that the velocity we observe macroscopically merely is the projection of all these numerous increments in the direction of motion. The DIST loop may coincide with the Compton oscillation, supporting the proposition that the Compton oscillation, which is associated with all particles, takes place in the scale of spacetime.

Since a displacement at the speed of light occurs instantaneously, the particle may be seen as being at two different locations in 4D spacetime at the same time. And, if the spatial displacement of each of these increments is comparable to the wavelength of the Compton oscillation it would be consistent with Heisenberg's uncertainty relation. However, in 5D hyperspace these two different locations would differ in the fifth dimension. This suggests the intriguing possibility that processes may exist in a 5D universe that involves influences beyond 4D spacetime.

Chess is a game that plays out in two dimensions. However, moving a piece makes use of the third dimension. Similarly the geometry of the world is four-dimensional but motion in this 4D world may take place via a fifth dimension. In 4D spacetime motion may seem mysterious since it involves instantaneous quantum jumps but in 5D hyperspace it becomes understandable because "instantaneous" change in location may take place via motion in the fifth dimension. It appears that the fifth dimension is not merely a piece of nice mathematics but might be as real as any of the four dimensions of spacetime.

This explanation may seem a bit speculative, but we cannot move beyond known physics without speculation. If it has merit it carries with it a hidden lesson; we must always remind ourselves that what we learn via our senses may not be enough to explain the world. Faulty premises and hidden realities may invalidate even our most reasonable conclusions.

Let us be humble; what we know is but a tiny fraction of what is to be known.

References

Anderson J. D., Laing P. A., Lau E. L., Anthony S. L., Nieto M. M., and. Turyshev S. G. "Study of the anomalous acceleration of Pioneer 10 and 11." *Physics Review Letters D* 65 (2003): 082004–54.

Bell, John S. *Speakable and Unspeakable in Quantum Mechanics.* Cambridge: Cambridge University Press, (1987).

Bohm, D. "A suggested interpretation of quantum theory in terms of 'hidden' variables, I and II." *Physics Review* 85 (1952): 166–93.

Bohm, D., and J. V. Vigier. "Model of the causal interpretation of quantum theory in terms of fluid with irregular fluctuations." *Physics Review* 96 (1954): 208–16.

Dingle, H. *Relativity for All.* Boston: Little &Brown, 1922.

Dingle, H. *The Special Theory of Relativity.* London: Meuthen, 1940.

Dingle, H. *Science at the Crossroads.* London: Martin, Brian & O'Keefe, 1972.

Dirac, P. A. M. "Long Range Forces and Broken Symmetries" *Proceedings of the Royal Society A: Mathematical, Physical & Engineering Sciences* 333 (1973): 403–18.

Djorgovski, S., and Spinrad H. "Toward the application of metric size function in galactic evolution and cosmology." *The Astrophysical Journal* 251 (1973): 417–23.

Dirac P.A.M. "The Quantum Theory of the Electron", *Proc. R. Soc. A (1928) vol. 117, no_778, 610-624*

Dürr J. S., Goldstein S., and Zanghi N. *Bohmian mechanics as the foundation of quantum mechanics.* In Bohmian mechanics and quantum theory, an appraisal. Dordrecht, the Netherlands: Kluwer Academic (1996). Ehrenfest's paradox. n.d. In *Wikipedia*

Ehrenfest Paradox: http://en.wikipedia.org/wiki/Ehrenfest%27s_paradox. (Aug. 11, 2011)

Einstein, A. "On the electrodynamics of moving bodies. *Annalen der Physik* 17 (1905): 891–921.

Einstein, A. "Kosmologische Betrachtungen zur allgemeinen Relativitätstheorie". *Königlich Preussische Akademie der Wissenschaften* VI. Berlin, (1917).

Einstein, A. "Die Feldgleichungen der Gravitation" (The field equations of gravitation). *Königlich Preussische Akademie der Wissenschaften:* (1915).844–47

Einstein, A. 'Über den Äther', *Verhandlungen der Schweizerischen Naturforschenden Gesellschaft* 105:2, 85–93. (1924)

Einstein, A. "Quantenmechanic und wirlichgkeit. Dialectica", 2, (1948):329–24.

Ernst, A. and Hsu J.-P. "First proposal of the universal speed of light by Voigt" 1887. *Chinese Journal of Physics* 30, (2001):211–29.

GALEX:www.seds.org/messier/more/m101_galex.html (Dec. 10, 2003)

Glanz, J. "Breakthrough of the year: Astronomy: Cosmic motion revealed." *Science* 282 (1998): 2156–57.

Hayasaka, Hideo, and Sakae Takeuchi. 1989. "Anomalous weight reduction on a gyroscope's right rotations around the vertical axis on the Earth." *Phys Physical Review Letters* 63 (25), (1998): 2697–2700.

Herbert Dingle. n.d. In *Wikipedia.* http://en.wikipedia.org/wiki/Herbert Dingle.

Holmes, Richard. *The age of wonder.* New York: Pantheon Books, (2008)

Horgan, John. *The end of science: Facing the limits of knowledge in the twilight of the scientific age.* Reading, MA: Helix Books. (1996)

Kepler, J. *The harmony of the world.* Trans E. J, Aiton, A. M. Duncan, and J. V. Field. Memoirs of the American Philosophical Society. Philadelphia: American Philosophical Society. (1697).

Kolesnik, Y. B. "Analysis of modern observations of the Sun and inner planets." *Astronomy and Astrophysics* 294, (1995): 876–94.

Kolesnik, Y. B. "Residual rotation of the FK5 system from optical observations of the Sun and planets 1960–1994." In *Dynamics, ephemerides and astrometry of the solar system*, ed. S. Ferraz-Mello, B. Morando, and J.-E. Arlot, 477–81. Dordrecht, the Netherlands: Reidel. (1996).

Kolesnik Y, and Masreliez C. J., Secular trends in the mean longitudes of the planets derived from optical observations, *AJ. 128, No. 2, p. 878, (2004)*

Krasinsky, G. A., Aleshkina E. Y., Pitijeva E. V., and Sveshnikov M. L. In *Relativity in celestial mechanics and astronomy*, ed. J. Kovalevsky and V. A. Brumberg, 215–XX. New York: Springer. (1986).

Langevin P. "L'évolution de l'espace et du temps." *Scientia* 10:31. (1911).

LaViolette, P. "Is the universe really expanding?" *The Astrophysical Journal* 301:544–53. (1986).

LaViolette, P. *Secrets of antigravity propulsion.* Rochester, VT: Bear & Co. (2008).

Lightman, A. *Ancient light.* Cambridge, MA: Harvard University Press. (1991).

Lubin, L.M., and Sandage A. "The Tolman surface brightness test for the reality of the expansion. IV. A measurement of the Tolman signal and the luminosity evolution of early-type galaxies." *Astrophysics* 122:1074–103. (2001).

Masreliez, C. J. "The scale expanding cosmos." *Astrophysics and Space Science* 336:399–447. (1999).

Masreliez, C. J. "Scale expanding cosmos theory I—An introduction." *Apeiron* 11 (3): 99–133. (2004a).

Masreliez, C. J. "Scale expanding cosmos theory II– Cosmic drag." *Apeiron* 11 (4): 1–29. (2004b).

Masreliez, C. J. "Scale expanding cosmos theory III—Gravitation." *Apeiron* 11 (4): 30–51. (2004c).

Masreliez, C. J. "Scale expanding cosmos theory IV—A possible link between general relativity and quantum mechanics." *Apeiron* 12 (1): 89–121. (2005a).

Masreliez, C. J. "A cosmological explanation to the Pioneer anomaly." *Astrophysics and Space Science* 299:83–108. (2005b).

Masreliez C. J. "On the origin of inertial force." *Apeiron* 13 (1): 43–77, (2006a)

Masreliez C. J. "The scale-expanding cosmos theory." *Nexus Magazine* 13 (June–July): 39-44. (2006b).

Masreliez, C. J. "Does cosmological scale-expansion explain the universe?" *Physics Essays* 19:91–122. (2006c).

Masreliez, C. J. "Motion, inertia and special relativity—a novel perspective." *Physica Scripta* 75:119-125. (2007a).

Masreliez, C. J. "Dynamic incremental scale transition with application to physics and cosmology." *Physica Scripta* 76:486–93. (2007b).

Masreliez, C. J. "Special relativity and inertia in curved spacetime." *Advanced Studies in Theoretical Physics* 2:795–815. (2008).

Masreliez, C. J. "Inertial field energy." *Advanced Studies in Theoretical Physics* 3:131–40. (2009).

Masreliez C. J., "Inertia and a fifth dimension—Special Relativity from a new Perspective", *Astrophys Space Sci* *326: 281–291,* (2010)

Metcalfe, N., T. Shanks, R. Fong, and Roche N. "Galaxy number counts—III. Deep CCD observations to B = 27.5 mag." *Monthly Notices of the Royal Astronomical Society* 273:257–76. (1995)

Michelson, A. A. & Morley, E. W. "On the Relative Motion of the Earth and the Luminiferous Ether". *American Journal of Science* 34: 333–345. (1887).

Neto, P. N., and P. I. Trajtenberg P. I. "On the localization of gravitational energy." *Brazilian Journal of Physics* 30:181–88. (2000)

Oesterwinter, C. and Cohen C. J. "New orbital elements for Moon and planets." *Celestial Mechanics* 5:317–95. (1972).

Reasenberg, R. D. and Shapire I. I. In *On the measurement of cosmological variation of the gravitational constant*, ed. L. Halpern, 71–XX. Gainesville: University Press of Florida. (1978).

REFERENCES

Perlmutter, S., Pennypacker C. R., G. Goldhaber, A. Goobar, R. A. Muller, H. J. M. Newberg, J. Desai, A. G. Kim, M. Y. Kim, I. A. Small, et al. "A type 1a supernova at $z = 0.347$." *Astrophysical Journal Letters* 440:L41–L45. (1995).

Perlmutter, S., Gabi S., Goldhaber G., Goobar A., Groom D. E., Hook I. M., Kim A. G., Kim M. Y., Lee J. C., Pain R., et al. "Measurements of the cosmological parameters Ω and from the first seven supernovae at $z \geq 0.35$." *The Astrophysics Journal* 483:565–81. (1997).

Perlmutter, S., Aldering G., Goldhaber G., Knop R. A., Nugent P.,. Castro P. G, Deustua S., Fabbro S., Goobar A., Groom D. E., et al. "Measurements of Ω and from 42 high-redshift supernovae." *The Astrophysical Journal* 517:565–86.

Perlmutter, S. 2003. "Supernovae, dark energy, and the accelerating universe." *Physics Today* (April): 53–60. (1999).

Poppe P. C. R., Leister N., Laclare F., and Delmas C. "Analysis of astrolabe measurements during 20 years: I. FK5 reference frame, personal and instrumental corrections." *Astronomical Journal* 116:2574–82. (1997).

Riess, A. G., Strolger L., Tonry J., Casertano S., Ferguson H. C., Mobasher B., Challis P., Filippenko A. V., Jha S., Weidong, Li, et al. "Type 1a supernova discoveries at $z < 1$ from the Hubble space telescope: Evidence for past deceleration and contraints on dark energy evolution." *The Astrophysical Journal* 607:665–87. (2004).

Schmidt B.P., Suntzeff N. B., Philips M. M., Schommer R. A., Clocchiatti A., Kirshnner R. P., Garnavich P., Challis P., Leibundgut B., Spyromilio J., et al.

"The High-Z Supernovae Search: Measuring cosmic deceleration and global curvature of the universe using type 1a supernovae." *The Astrophysical Journal* 507:46–63. (1998).

Schwarzchild, K. "Über das gravitationsfeld eins massenpunkets nach der Einsteinschen theorie." *Sitzungsberichte der Königlich Preussischen Akademie der Wissenschaften* 1:189–96. (1916).

Seidelman P. K., Santoro E. J., and Pulkkinen K. F. In *Dynamical astronomy,* ed. V. Szebenhey and B. Balazs, 55–XX. Austin: University of Texas Press. (1985).

Seidelman P. K., Santoro E. J., Pulkkinen K. F. . In *Relativity in celestial mechanics and astrometry*, J. Kovalevsky and V. A. Brumberg, 89–XX. Dordrecht, the Netherlands: Kluwer. (1986).

Smolin, L. *The trouble with physics.* New York: Houghton Mifflin. (2006).

Standish E. M., Williams J.G., 1990, in Lieske J.H. and Abalakin V.K. (Eds.) *Inertial Coordinate System on the Sky*, Kluwer, Dordrecht, p.173

Standish E. M. "Time scales in JPL and CfA ephemerides." *Astronomy & Astrophysics* 336: 381–84. (1998).

Tolman, R. C. *Relativity, thermodynamics and cosmology.* New York: Dover Publications IncValone, Thomas. 2004. *Electro Gravitics II.* Washington, DC: Integrity Research Institute. (1987).

Turyshev S. G. et al, http://arxiv.org/PS_cache/arxiv/pdf/1107/1107.2886v1.pdf (2011)

Vassilatos, G. *Secrets of cold war technology*. Kempton, IL: Adventures Unlimited Press. (2000).

Voigt, W., *Göttinger Nachrichten* 7: 41-51 (1887)

Wells, J., "Coral growth and geochronometry." *Nature, 197,* (1963): 948–950

Weyl, H. K. H. *Space—Time—Matter*. Trans. H. A. Brose. London: Methun & Co. (1921).

Woit, P. *Not even wrong.* New York: Basic Books. (2006).

Yao, Z. and Smith C. In *Mapping the sky*, ed. S. Débarbat, J. A. Eddy, H. K. Eichhorn, and A. R. Upgren, XX–XX. Dordrecht, the Netherlands: Kluwer. (1988).

Yao, Z. and Smith C. "The effect of various solar ephemerides on equator and equinox solutions from observations of the sun with the reversible transit at Cape observatory from 1907 to 1959." *Astrophysics and Space Science* 177:181–88. (1991).

Yao, Z. and Smith C. 1993. "Equator and equinox solutions from Meridian Circle observations of the Sun, Mercury and Venus at the Cape of Good Hope and the U.S. Naval Observatory from 1907 to 1971.: In *Developments in astrometry and their impact on astrophysics and geodynamics*, ed. I. I. Muller and B. Kolaczek, 403. Dordrecht, the Netherlands: Kluwer.

Sofue Y. and Rubin V. "Rotation curves of spiral galaxies. Annual Review of Astronomy and Astrophysics" 39:137–74. (2001).

Zero-point energy. n.d. In *Wikipedia*. http://en.wikipedia.org/wiki/Zero-point_energy.

Zhuck, N. A., et al. 2001. "Quasars and the large-scale structure of the Universe." *Spacetime & Science* 2 (5): XX–XX.

INDEX